Transformers: Basics, Maintenance, and Diagnostics

.S. Department of the Interior
ureau of Reclamation
echnical Service Center
frastructure Services Division
ydroelectric Research and Technical Services Group
enver, Colorado

April 2005

For sale by the Superintendent of Documents, Government Printing Office
Internet: bookstore.gpo.gov • Phone: toll free (866) 512-1800; • DC area (201) 512-1800
• Fax: (202) 512-2250 • Mail: Stop SSOP, Washington DC 20402-0001
ISBN 0-16-074945-X
S/N 024-003-00192-9

PREFACE

Transformers have been used at powerplants since the inception of alternating-current generation, a century ago. While operating principles of transformers remain the same, the challenges of maintaining and testing transformers have evolved along with transformer design and construction. Modern transformers are designed to closer tolerances than transformers in the past. Thus, effective, regular maintenance and testing is even more essential to continued operation when traditional "overdesign" cannot be relied on to overcome abnormal conditions. The utility engineer must be familiar with all aspects of maintenance and testing and make use of state-of-the-art tools and techniques for evaluating transformer condition. While on-line diagnostic systems and computerized testing methods are very helpful, they are not a substitute for sound engineering judgment and expertise.

This volume provides timely, practical advice to those seeking to better understand how transformers work, how they are best maintained, and how to test and evaluate their condition. It has been developed with the assistance of Bureau of Reclamation engineers responsible for operating and maintaining transformers at important powerplants in the Western States. Support and funding was provided through the Reclamation Power Resources Office in Denver and via the Manuals and Standards development program.

The authors gratefully acknowledge the assistance of all who contributed.

Hydroelectric Research and
Technical Services Group
Denver, Colorado
April 2005

Contents

Contents (continued)

Contents (continued)

Contents (continued)

Contents (continued)

Contents (continued)

Contents (continued)

Tables

Tables (continued)

Figures

Figures (continued)

Figures (continued)

1. Introduction

This document was created to provide guidance to Bureau of Reclamation (Reclamation) powerplant personnel in maintenance, diagnostics, and testing of transformers and associated equipment.

This document applies primarily to the maintenance and diagnostics of oil-filled power transformers (500 kilovoltamperes [kVA] and larger), owned and operated by Reclamation, although routine maintenance of other transformer types is addressed as well. Specific technical details are included in other documents and are referenced in this document.

Guidance and recommendations herein are based on industry standards and experience gained at Reclamation facilities. However, equipment and situations vary greatly, and sound engineering and management judgment must be exercised when applying these diagnostics. All available information must be considered (e.g., manufacturer's and transformer experts' recommendations, unusual operating conditions, personal experience with the equipment, etc.) in conjunction with this document.

2. Introduction to Transformers

Generator step-up (GSU) transformers represent the second largest capital investment in Reclamation power production—second only to generators. Reclamation has hundreds, perhaps thousands, of transformers, in addition to hundreds of large GSU transformers. Reclamation has transformers as small as a camera battery charger, about one-half the size of a coffee cup, to huge generator step-up transformers near the size of a small house. The total investment in transformers may well exceed generator investment. Transformers are extremely important to Reclamation, and it is necessary to understand their basic functions.

A transformer has no internal moving parts, and it transfers energy from one circuit to another by electromagnetic induction. External cooling may include heat exchangers, radiators, fans, and oil pumps. Radiators and fans are evident in figure 1. The large horizontal tank at the top is a conservator. Transformers are typically used because a change in voltage is needed. Power transformers are defined as transformers rated 500 kVA and larger. Larger transformers are oil-filled for insulation and cooling; a typical GSU transformer may contain several thousand gallons of oil. One must always be aware of the possibility of spills, leaks, fires, and environmental risks this oil poses.

Figure 1 – Typical GSU Three-Phase Transformer.

Transformers smaller than 500 kVA are generally called distribution transformers. Pole-top and small, pad-mounted transformers that serve residences and small businesses are typically distribution transformers. Generator step-up transformers, used in Reclamation powerplants, receive electrical energy at generator voltage and increase it to a higher voltage for transmission lines. Conversely, a step-down transformer receives energy at a higher voltage and delivers it at a lower voltage for distribution to various loads.

All electrical devices using coils (in this case, transformers) are constant wattage devices. This means voltage multiplied by current must remain constant; therefore, when voltage is "stepped-up," the current is "stepped-down" (and vice versa). Transformers transfer electrical energy between circuits completely insulated from each other. This makes it possible to use very high (stepped-up) voltages for transmission lines, resulting in a lower (stepped-down) current. Higher voltage and lower current reduce the required size and cost of transmission lines and reduce transmission losses as well. Transformers have made possible economic delivery of electric power over long distances.

Transformers do not require as much attention as most other equipment; however, the care and maintenance they do require is absolutely critical. Because of their reliability, maintenance is sometimes ignored, causing reduced service life and, at times, outright failure.

2.1 Principle of Operation

Transformer function is based on the principle that electrical energy is transferred efficiently by magnetic induction from one circuit to another. When one winding of a transformer is energized from an alternating current (AC) source, an alternating magnetic field is established in the transformer core. Alternating magnetic lines of force, called "flux," circulate through the core. With a second winding around the same core, a voltage is induced by the alternating flux lines. A circuit, connected to the terminals of the second winding, results in current flow.

Each phase of a transformer is composed of two separate coil windings wound on a common core. The low-voltage winding is placed nearest the core; the high-voltage winding is then placed around both the low-voltage winding and core. See figure 2 which shows internal construction of one phase. The core is typically made from very thin steel laminations, each coated with insulation. By insulating between individual laminations, losses are reduced. The steel core provides a low resistance path for magnetic flux. Both high- and low-voltage windings are insulated from the core and from each other, and leads

LAMINATED MAGNETIC STEEL CORE

HIGH DENSITY WOOD OR PAPER WINDING STICKS

HIGH DENSITY CRAFT PAPER TUBE BETWEEN PRIMARY AND SECONDARY WINDING STICKS

HIGH DENSITY WOOD OR PAPER WINDING STICKS

HIGH VOLTAGE WINDING

PAPER INSULATION

PHASE A   PHASE B   PHASE C

HIGH DENSITY CRAFT PAPER TUBE

COPPER

PAPER INSULATION

LOW VOLTAGE WINDING

HEAVY CELLULOSE PHASE TO PHASE INSULATION

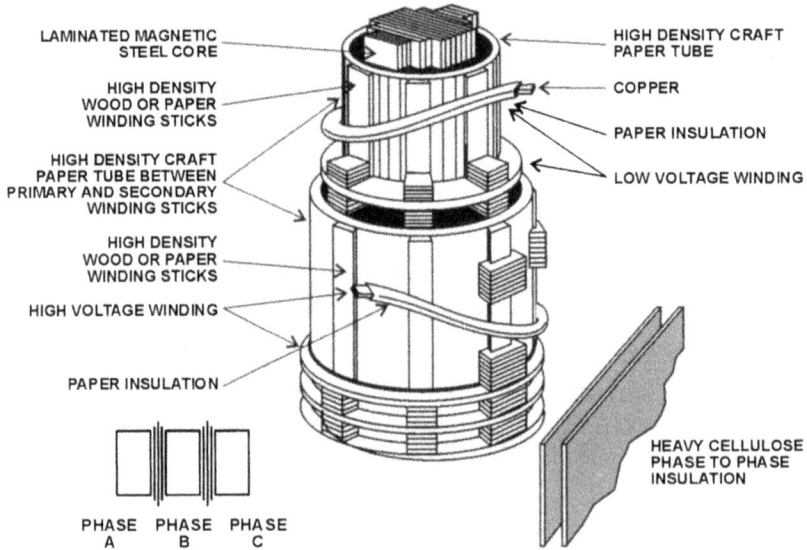

Figure 2 – Transformer Construction.

are brought out through insulating bushings. A three-phase transformer typically has a core with three legs and has both high-voltage and low-voltage windings around each leg. Special paper and wood are used for insulation and internal structural support.

2.2 Transformer Action

Transformer action depends upon magnetic lines of force (flux) mentioned above. At the instant a transformer primary is energized with AC, a flow of electrons (current) begins. During the instant of switch closing, buildup of current and magnetic field occurs. As current begins the positive portion of the sine wave, lines of magnetic force (flux) develop outward from the coil and continue to expand until the current is at its positive peak. The magnetic field is also at its positive peak. The current sine wave then begins to decrease, crosses zero, and goes negative until it reaches its negative peak. The magnetic flux switches direction and also reaches its peak in the opposite direction. With an AC power circuit, the current changes (alternates) continually 60 times per second, which is standard in the United States. Other countries may use other frequencies. In Europe, 50 cycles per second is common.

Strength of a magnetic field depends on the amount of current and number of turns in the winding. When current is reduced, the magnetic field shrinks. When the current is switched off, the magnetic field collapses.

When a coil is placed in an AC circuit, as shown in figure 3, current in the primary coil will be accompanied by a constantly rising and collapsing magnetic field. When another coil is placed within the alternating magnetic field of the first coil, the rising and collapsing flux will induce voltage in the second coil.

When an external circuit is connected to the second coil, the induced voltage in the coil will cause a current in the second coil. The coils are said to be magnetically coupled; they are, however, electrically isolated from each other.

Many transformers have separate coils, as shown in figure 3, and contain many turns of wire and a magnetic core, which forms a path for and concentrates the magnetic flux. The winding receiving electrical energy from the source is called the primary winding. The winding which receives energy from the primary winding, via the magnetic field, is called the "secondary" winding.

Either the high- or low-voltage winding can be the primary or the secondary. With GSUs at Reclamation powerplants, the primary winding is the low-voltage side (generator voltage), and the high-voltage side is the secondary winding (transmission voltage). Where power is used (i.e., at residences or businesses), the primary winding is the high-voltage side, and the secondary winding is the low-voltage side.

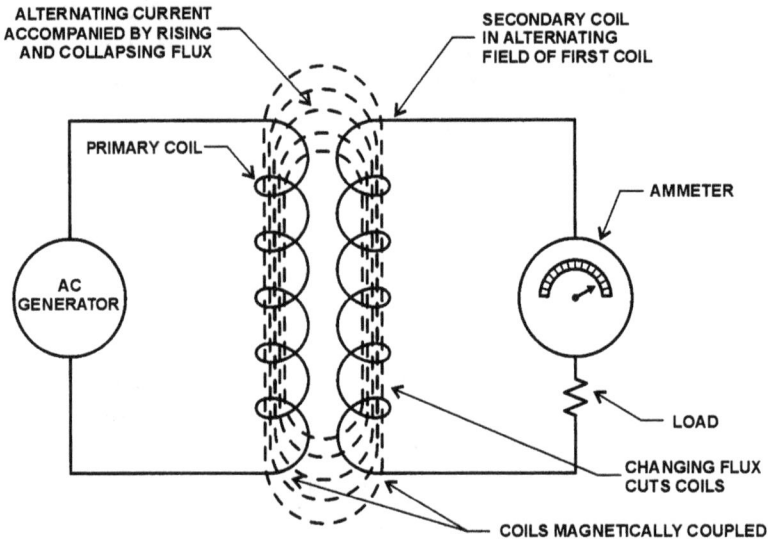

Figure 3 – Transformer Action.

The amount of voltage induced in each turn of the secondary winding will be the same as the voltage across each turn of the primary winding. The total amount of voltage induced will be equal to the sum of the voltages induced in each turn. Therefore, if the secondary winding has more turns than the primary, a greater voltage will be induced in the secondary; and the transformer is known as a step-up transformer. If the secondary winding has fewer turns than the primary, a lower voltage will be induced in the secondary; and the transformer is a step-down transformer. Note that the primary is always connected to the source of power, and the secondary is always connected to the load.

In actual practice, the amount of power available from the secondary will be slightly less than the amount supplied to the primary because of losses in the transformer itself.

When an AC generator is connected to the primary coil of a transformer (figure 4), electrons flow through the coil due to the

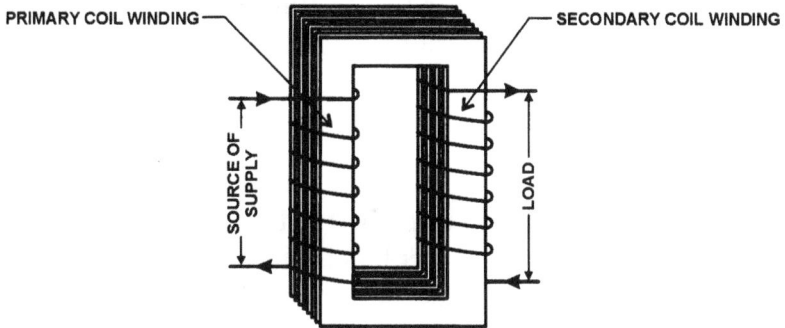

Figure 4 – Transformer.

generator voltage. Alternating current varies, and accompanying magnetic flux varies, cutting both transformer coils and inducing voltage in each coil circuit.

The voltage induced in the primary circuit opposes the applied voltage and is known as back voltage or back electro-motive-force (back EMF). When the secondary circuit is open, back EMF, along with the primary circuit resistance, acts to limit the primary current. Primary current must be sufficient to maintain enough magnetic field to produce the required back EMF.

When the secondary circuit is closed and a load is applied, current appears in the secondary due to induced voltage, resulting from flux created by the primary current. This secondary current sets up a second magnetic field in the transformer in the opposite direction of the primary field. Thus, the two fields oppose each other and result in a combined magnetic field of less strength than the single field produced by the primary with the secondary open. This reduces the back voltage (back EMF) of the primary and causes the primary current to increase. The primary current increases until it re-establishes the total magnetic field at its original strength.

7

In transformers, a balanced condition must always exist between the primary and secondary magnetic fields. Volts times amperes (amps) must also be balanced (be the same) on both primary and secondary. The required primary voltage and current must be supplied to maintain the transformer losses and secondary load.

2.3 Transformer Voltage and Current

If the small amount of transformer loss is ignored, the back-voltage (back EMF) of the primary must equal the applied voltage. The magnetic field, which induces the back-voltage in the primary, also cuts the secondary coil. If the secondary coil has the same number of turns as the primary, the voltage induced in the secondary will equal the back-voltage induced in the primary (or the applied voltage). If the secondary coil has twice as many turns as the primary, it will be cut twice as many times by the flux, and twice the applied primary voltage will be induced in the secondary. The total induced voltage in each winding is proportional to the number of turns in that winding. If E_1 is the primary voltage and I_1 the primary current, E_2 the secondary voltage and I_2 the secondary current, N_1 the primary turns and N_2 the secondary turns, then:

$$\frac{E_1}{E_2} = \frac{N_1}{N_2} = \frac{I_2}{I_1}$$

Note that the current is inversely proportional to both voltage and number of turns. This means (as discussed earlier) that if voltage is stepped up, the current must be stepped down and vice versa. The number of turns remains constant unless there is a tap changer (discussed later).

The power output or input of a transformer equals volts times amperes (E x I). If the small amount of transformer loss is disregarded, input equals output or:

$$E_1 \times I_1 = E_2 \times I_2$$

If the primary voltage of a transformer is 110 volts (V), the primary winding has 100 turns, and the secondary winding has 400 turns, what will the secondary voltage be?

$$\frac{E_1}{E_2} = \frac{N_1}{N_2} \qquad \frac{110}{E_2} = \frac{100}{400}$$

$$100\ E_2 = 44{,}000 \qquad E_2 = 440 \text{ volts}$$

If the primary current is 20 amps, what will the secondary current be?

$$E_2 \times I_2 = E_1 \times I_1$$

$$440 \times I_2 = 110 \times 20 = 2{,}200$$

$$I_2 = 5 \text{ amps}$$

Since there is a ratio of 1 to 4 between the turns in the primary and secondary circuits, there must be a ratio of 1 to 4 between the primary and secondary voltage and a ratio of 4 to 1 between the primary and secondary current. As voltage is stepped up, the current is stepped down, keeping volts multiplied by amps constant. This is referred to as "volt amps."

As mentioned earlier and further illustrated in figure 5, when the number of turns or voltage on the secondary of a transformer is greater than that of the primary, it is known as a step-up transformer. When the number of turns or voltage on the secondary is less than that of the primary, it is known as a step-down transformer. A power transformer used to tie two systems together may feed current either way between systems, or act as a step-up or step-down transformer, depending on where power is being generated and where it is consumed. As mentioned above, either winding could be the primary or secondary. To eliminate this confusion, in power generation, windings of transformers are often referred to as high-side and low-side windings, depending on the relative values of the voltages.

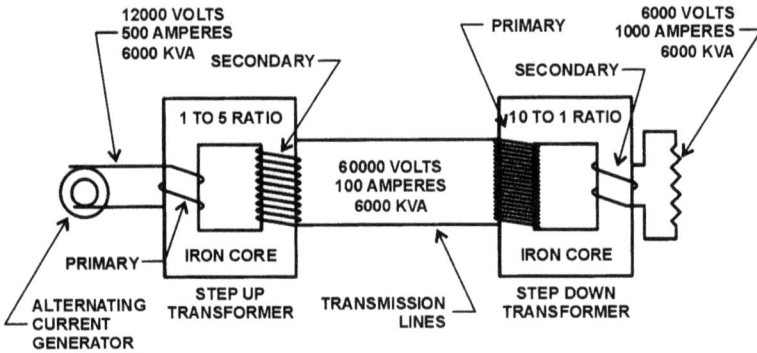

Figure 5 – Step-Up and Step-Down Transformers.

Note that kVA (amps times volts) remains constant throughout the above circuit on both sides of each transformer, which is why they are called constant wattage devices.

Efficiencies of well-designed power transformers are very high, averaging over 98 percent (%). The only losses are due to core losses, maintaining the alternating magnetic field, resistance losses in the coils, and power used for cooling. The main reason for high efficiencies of power transformers, compared to other equipment, is the absence of moving parts. Transformers are called static AC machines.

2.4 The Magnetic Circuit

A magnetic circuit or core of a transformer is designed to provide a path for the magnetic field, which is necessary for induction of voltages between windings. A path of low reluctance (i.e., resistance to magnetic lines of force), consisting of thin silicon, sheet steel laminations, is used for this purpose. In addition to providing a low reluctance path for the magnetic field, the core is designed to prevent circulating electric currents within the steel itself. Circulating currents, called eddy currents, cause heating and energy loss. They are due to voltages induced in the steel of the core, which is constantly subject to alternating magnetic fields. Steel itself is a conductor, and changing lines of magnetic flux also induce a voltage and current in this conductor. By using very thin sheets of steel with insulating material between sheets, eddy currents (losses) are greatly reduced.

The two common arrangements of the magnetic path and the windings are shown in figure 6 and 7. In the core-type (core form) transformer, the windings surround the core.

A section of both primary and secondary windings are wound on each leg of the core, the low voltage winding is wound next to the core, and the high voltage winding is wound over the low voltage.

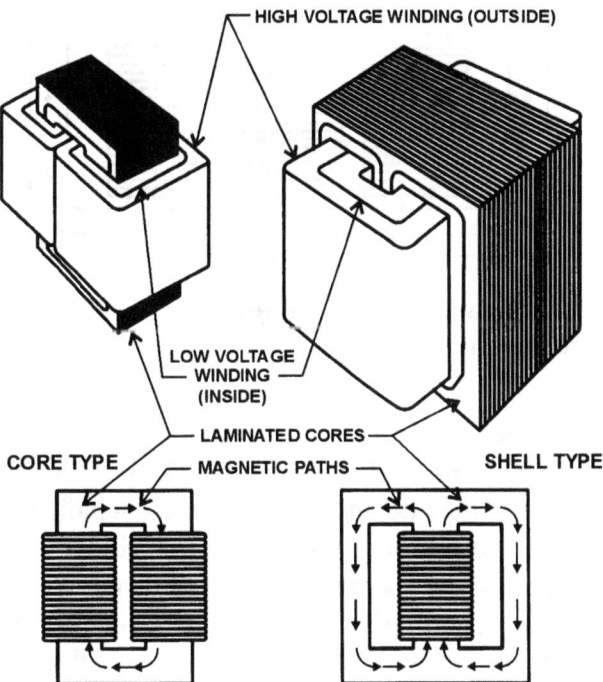

Figure 6 – Magnetic Circuits.

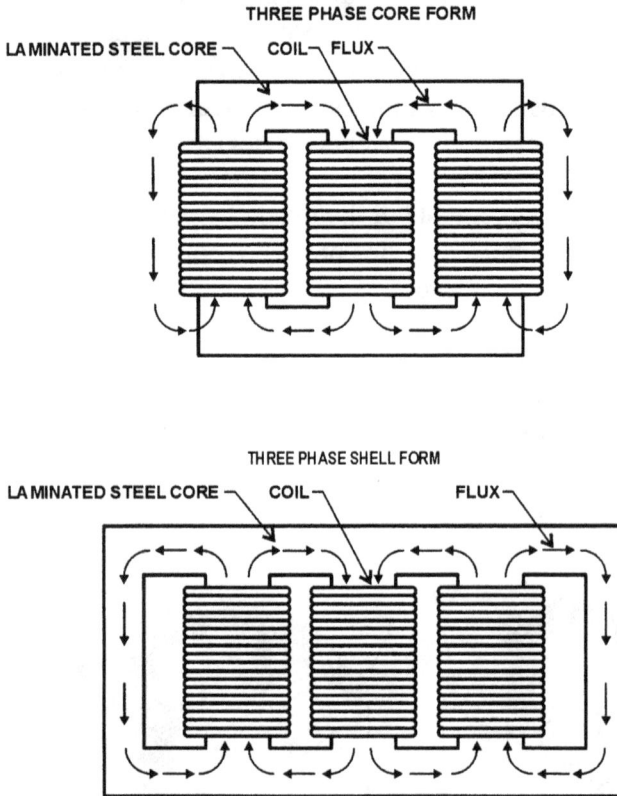

Figure 7 – Three-Phase Core Form and Three-Phase Shell
Form Transformer Units.

In a shell-type (shell form) transformer, the steel magnetic circuit
(core) forms a shell surrounding the windings. In a core form, the
windings are on the outside; in a shell form, the windings are on the
inside. In power transformers, the electrical windings are arranged so
that practically all of the magnetic lines of force go through both the
primary and secondary windings. A small percentage of the magnetic
lines of force goes outside the core, and this is called leakage flux.
Larger transformers, such as Reclamation GSU transformers, are
almost always shell type.

Note that, in the shell form transformers, (see figure 7) the magnetic flux, external to the coils on both left and right extremes, has complete magnetic paths for stray and zero sequence flux to return to the coils. In the core form, it can easily be seen from the figure that, on both left and right extremes, there are no return paths. This means that the flux must use external tank walls and the insulating medium for return paths. This increases core losses and decreases overall efficiency and shows why most large transformers are built as shell form units.

2.5 Core Losses

Since magnetic lines of force in a transformer are constantly changing in value and direction, heat is developed because of the hysteresis of the magnetic material (friction of the molecules). This heat must be removed; therefore, it represents an energy loss of the transformer. High temperatures in a transformer will drastically shorten the life of insulating materials used in the windings and structures. For every 8 degrees Celsius (°C) temperature rise, life of the transformer is cut by one-half; therefore, maintenance of cooling systems is critical.

Losses of energy, which appears as heat due both to hysteresis and to eddy currents in the magnetic path, is known as core losses. Since these losses are due to alternating magnetic fields, they occur in a transformer whenever the primary is energized, even though no load is on the secondary winding.

2.6 Copper Losses

There is some loss of energy in a transformer due to resistance of the primary winding to the magnetizing current, even when no load is connected to the transformer. This loss appears as heat generated in the winding and must also be removed by the cooling system. When a load is connected to a transformer and electrical currents exist in both primary and secondary windings, further losses of electrical energy occur. These losses, due to resistance of the windings, are called copper losses (or the I^2R losses).

2.7 Transformer Rating

Capacity (or rating) of a transformer is limited by the temperature that the insulation can tolerate. Ratings can be increased by reducing core and copper losses, by increasing the rate of heat dissipation (better cooling), or by improving transformer insulation so it will withstand higher temperatures. A physically larger transformer can dissipate more heat, due to the increased area and increased volume of oil. A transformer is only as strong as its weakest link, and the weakest link is the paper insulation, which begins to degrade around 100 °C. This means that a transformer must be operated with the "hottest spot" cooler than this degradation temperature, or service life is greatly reduced. Reclamation typically orders transformers larger than required, which aids in heat removal and increases transformer life.

Ratings of transformers are obtained by simply multiplying the current times the voltage. Small transformers are rated in "VA," volts times amperes. As size increases, 1 kilovoltampere (kVA) means 1,000 voltamperes, 1 megavoltampere (MVA) means 1 million voltamperes. Large GSUs may be rated in hundreds of MVAs. A GSU transformer can cost well over a million dollars and take 18 months to 2 years or longer to obtain. Each one is designed for a specific application. If one fails, this may mean a unit or whole plant could be down for as long 2 years, costing multiple millions of dollars in lost generation, in addition to the replacement cost of the transformer itself. This is one reason that proper maintenance is critical.

2.8 Percent Impedance

The impedance of a transformer is the total opposition offered an alternating current. This may be calculated for each winding. However, a rather simple test provides a practical method of measuring the equivalent impedance of a transformer without separating the impedance of the windings. When referring to impedance of a transformer, it is the equivalent impedance that is meant. In order to determine equivalent impedance, one winding of the transformer is short circuited, and just enough voltage is applied to the other winding to create full load current to flow in the short circuited winding. This voltage is known as the impedance voltage. Either winding may be short-circuited for this test, but it is usually more convenient to short circuit the low-voltage winding. The transformer impedance value is given on the nameplate in percent. This means that the voltage drop due to the impedance is expressed as a percent of rated voltage. For example, if a 2,400/240-volt transformer has a measured impedance voltage of 72 volts on the high-voltage windings, its impedance (Z), expressed as a percent, is:

$$\text{percent } Z = \frac{72}{2,400} \text{ x } 100 = 3 \text{ percent}$$

This means there would be a 72-volt drop in the high-voltage winding at full load due to losses in the windings and core. Only 1 or 2% of the losses are due to the core; about 98% are due to the winding impedance. If the transformer were not operating at full load, the voltage drop would be less. If an actual impedance value in ohms is needed for the high-voltage side:

$$Z = \frac{V}{I}$$

where V is the voltage drop or, in this case, 72 volts; and I is the full load current in the primary winding. If the full load current is 10 amps:

$$Z = \frac{72 V}{10 a} = 7.2 \text{ ohms}$$

Of course, one must remember that impedance is made up of both resistive and reactive components.

2.9 Internal Forces

During normal operation, internal structures and windings are subjected to mechanical forces due to the magnetic forces. These forces are illustrated in figure 8. By designing the internal structure very strong to withstand these forces over a long period of time, service life can be extended. However, in a large transformer during a "through fault" (fault current passing through a transformer), forces can reach millions of pounds, pulling the coils up and down and pulling them apart 60 times per second. Notice in figure 8 that the internal low-voltage coil is being pulled downward, while the high-voltage winding is pulled up, in the opposite direction. At the same time, the right-hand part of the figure shows that the high- and low-voltage coils are being forced apart. Keep in mind that these forces are reversing 60 times each second. It is obvious why internal structures of transformers must be built incredibly strong.

Many times, if fault currents are high, these forces can rip a transformer apart and cause electrical faults inside the transformer itself. This normally results in arcing inside the transformer that can result in explosive failure of the tank, throwing flaming oil over a wide

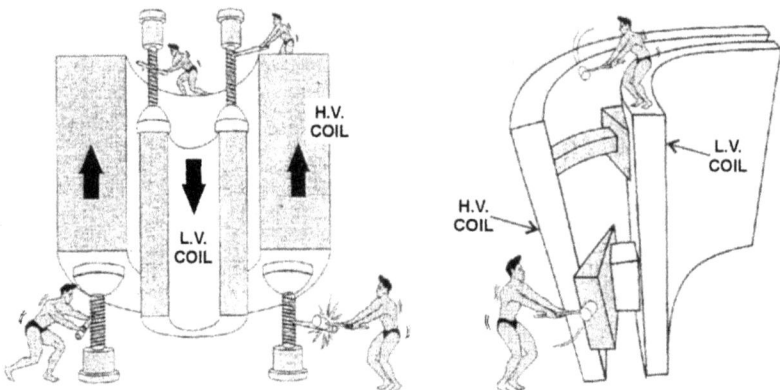

Figure 8 – Transformer Internal Forces.

area. There are protective relaying systems to protect against this possibility, although explosive failures do occur occasionally.

2.10 Autotransformers

It is possible to obtain transformer action by means of a single coil, provided that there is a "tap connection" somewhere along the winding. Transformers having only one winding are called autotransformers, shown schematically in figure 9.

An autotransformer has the usual magnetic core but only one winding, which is common to both the primary and secondary circuits.

The primary is always the portion of the winding connected to the AC power source. This transformer may be used to step voltage up or down. If the primary is the total winding and is connected to a supply, and the secondary circuit is connected across only a portion of the winding (as shown), the secondary voltage is "stepped-down."

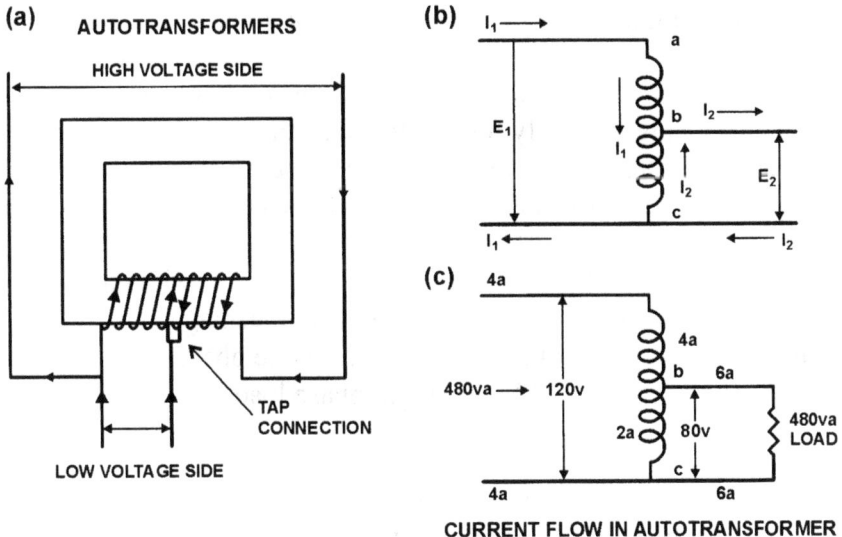

Figure 9 – Autotransformers.

17

If only a portion of the winding is the primary and is connected to the supply voltage and the secondary includes all the winding, then the voltage will be "stepped-up" in proportion to the ratio of the total turns to the number of connected turns in the primary winding.

When primary current I_1 is in the direction of the arrow, secondary current, I_2, is in the opposite direction, as in figure 9b. Therefore, in the portion of the winding between points b and c, current is the difference of I_1 and I_2. If the requirement is to step the voltage up (or down) only a small amount, then the transformer ratio is small—E_1 and E_2 are nearly equal. Currents I_1 and I_2 are also nearly equal. The portion of the winding between b and c, which carries the difference of the currents, can be made of a much smaller conductor, since the current is much lower.

Under these circumstances, the autotransformer is much cheaper than the two-coil transformer of the same rating. However, the disadvantage of the autotransformer is that the primary and secondary circuits are electrically connected and, therefore, could not safely be used for stepping down from high voltage to a voltage suitable for plant loads. The autotransformer, however, is extensively used for reducing line voltage for step increases in starting larger induction motors. There are generally four or five taps that are changed by timers so that more of the winding is added in each step until the full voltage is applied across the motor. This avoids the large inrush current required when starting motors at full line voltage. This transformer is also extensively used for "buck-boost" when the voltage needs to be stepped up or down only a small percentage. One very common example is boosting 208 V up from one phase of a 120/208-V three-phase system, to 220 V for single-phase loads.

2.11 Instrument Transformers

Instrument transformers (figure 10) are used for measuring and control purposes. They provide currents and voltages proportional to the primary, but there is less danger to instruments and personnel.

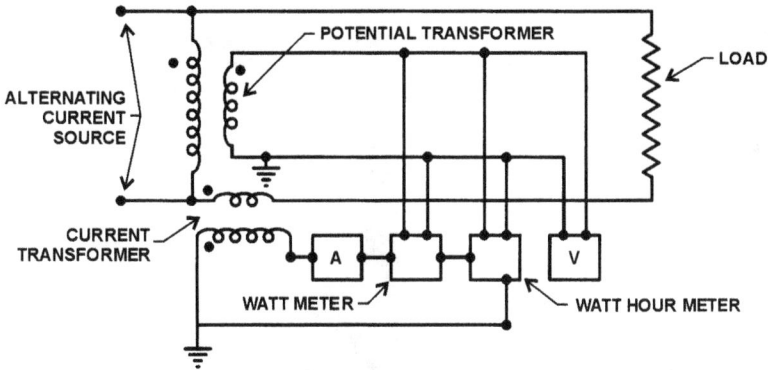

Figure 10 – Connections of Instrument Transformers.

Those transformers used to step voltage down are known as potential transformers (PTs) and those used to step current down are known as current transformers (CTs).

The function of a PT is to accurately measure voltage on the primary, while a CT is used to measure current on the primary.

2.12 Potential Transformers

Potential transformers (figure 11) are used with voltmeters, wattmeters, watt-hour meters, power-factor meters, frequency meters, synchroscopes and synchronizing apparatus, protective and regulating relays, and undervoltage and overvoltage trip coils of circuit breakers. One potential transformer can be used for a number of instruments if the total current required by the instruments connected to the secondary winding does not exceed the transformer rating.

Potential transformers are usually rated 50 to 200 volt-amperes at 120 secondary volts. The secondary terminals should never be short circuited because a heavy current will result, which can damage the windings.

Potential Transformer in switchyard.

Inside Potential Transformer with Fuses.

Figure 11 – Potential Transformers.

2.13 Current Transformers

The primary of a current transformer typically has only one turn. This is not really a turn or wrap around the core but just a conductor or bus going through the "window." The primary never has more than a very few turns, while the secondary may have a great many turns, depending upon how much the current must be stepped down. In most cases, the primary of a current transformer is a single wire or bus bar, and the secondary is wound on a laminated magnetic core, placed around the conductor in which the current needs to be measured, as illustrated in figure 12.

If primary current exists and the secondary circuit of a CT is closed, the winding builds and maintains a counter or back EMF to the primary magnetizing force. Should the secondary be opened with current in the primary, the counter EMF is removed; and the primary magnetizing force builds up an extremely high

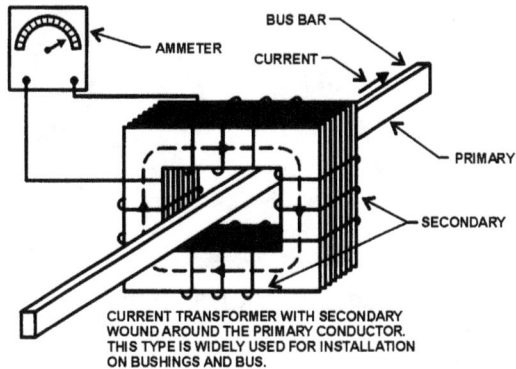

CURRENT TRANSFORMER WITH SECONDARY
WOUND AROUND THE PRIMARY CONDUCTOR.
THIS TYPE IS WIDELY USED FOR INSTALLATION
ON BUSHINGS AND BUS.

Figure 12 – Current Transformer.

voltage in the secondary, which is dangerous to personnel and can destroy the current transformer.

CAUTION:

For this reason, the secondary of a current transformer should always be shorted before removing a relay from its case or removing any other device that the CT operates. This protects the CT from overvoltage.

Current transformers are used with ammeters, wattmeters, power-factor meters, watt-hour meters, compensators, protective and regulating relays, and trip coils of circuit breakers. One CT can be used to operate several instruments, provided the combined loads of the instruments do not exceed that for which the CT is rated. Secondary windings are usually rated at 5 amperes. A variety of current transformers are shown in figure 13. Many times, CTs have several taps on the secondary winding to adjust the range of current possible to measure on the primary.

Lab Current Transformer.

Current Transformer in
Switchyard.

Current
Transformer in
Figure 13 – Photograph of Current Transformers.   Switchyard.

2.14 Transformer Taps

Most power transformers have taps on either primary or secondary
windings to vary the number of turns and, thus, the output voltage.
The percentage of voltage change, above or below normal, between
different tap positions varies in different transformers. In oil-cooled
transformers, tap leads are brought to a tap changer, located beneath
the oil inside the tank, or brought to an oil-filled tap changer,
externally located. Taps on dry-type transformers are brought to
insulated terminal boards located inside the metal housing, accessible
by removing a panel.

Some transformers taps can be changed under load, while other
transformers must be de-energized. When it is necessary to change
taps frequently to meet changing conditions, taps that can be changed
under load are used. This is accomplished by means of a motor that
may be controlled either manually or automatically. Automatic
operation is achieved by changing taps to maintain constant voltage as
system conditions change. A common range of adjustment is plus or

minus 10%. At Reclamation powerplants, de-energized tap changers (DETC) are used and can only be changed with the transformer off-line. A very few load tap changers (LTC) are used at Grand Coulee between the 500-kilovolt (kV) (volts x 1,000) and 220-kV switchyards.

A bypass device is sometimes used across tap changers to ensure power flow in case of contact failure. This prevents failure of the transformer in case excessive voltage appears across faulty contacts.

2.15 Transformer Bushings
The two most common types of bushings used on transformers as main lead entrances are solid porcelain bushings on smaller transformers and oil-filled condenser bushings on larger transformers.

Solid porcelain bushings consist of high-grade porcelain cylinders that conductors pass through. Outside surfaces have a series of skirts to increase the leakage path distance to the grounded metal case. High-voltage bushings are generally oil-filled condenser type. Condenser types have a central conductor wound with alternating layers of paper insulation and tin foil and filled with insulating oil. This results in a path from the conductor to the grounded tank, consisting of a series of condensers. The layers are designed to provide approximately equal voltage drops between each condenser layer.

Acceptance and routine maintenance tests most often used for checking the condition of bushings are Doble power factor tests. The power factor of a bushing in good condition will remain relatively stable throughout the service life. A good indication of insulation deterioration is a slowly rising power factor. The most common cause of failure is moisture entrance through the top bushing seal. This condition will be revealed before failure by routine Doble testing. If Doble testing is not performed regularly, explosive failure is the eventual result of a leaking bushing. This, many times, results in a catastrophic and expensive failure of the transformer as well.

2.16 Transformer Polarity

With power or distribution transformers, polarity is important only if the need arises to parallel transformers to gain additional capacity or to hook up three single-phase transformers to make a three-phase bank. The way the connections are made affects angular displacement, phase rotation, and direction of rotation of connected motors. Polarity is also important when hooking up current transformers for relay protection and metering. Transformer polarity depends on which direction coils are wound around the core (clockwise or counterclockwise) and how the leads are brought out. Transformers are sometimes marked at their terminals with polarity marks. Often, polarity marks are shown as white paint dots (for plus) or plus-minus marks on the transformer and symbols on the nameplate. These marks show the connections where the input and output voltages (and currents) have the same instantaneous polarity.

More often, transformer polarity is shown simply by the American National Standards Institute (ANSI) designations of the winding leads as H_1, H_2 and X_1, X_2. By ANSI standards, if you face the low-voltage side of a single-phase transformer (the side marked X_1, X_2), the H_1 connection will always be on your far left. See the single-phase diagrams in figure 14. If the terminal marked X_1 is also on your left, it is subtractive polarity. If the X_1 terminal is on your right, it is additive polarity. Additive polarity is common for small distribution transformers. Large transformers, such as GSUs at Reclamation powerplants, are generally subtractive polarity.

It is also helpful to think of polarity marks in terms of current direction. At any instant when the current direction is into a polarity marked terminal of the primary winding, the current direction is out of the terminal with the same polarity mark in the secondary winding. It is the same as if there were a continuous circuit across the two windings.

Polarity is a convenient way of stating how leads are brought out. If you want to test for polarity, connect the transformer as shown

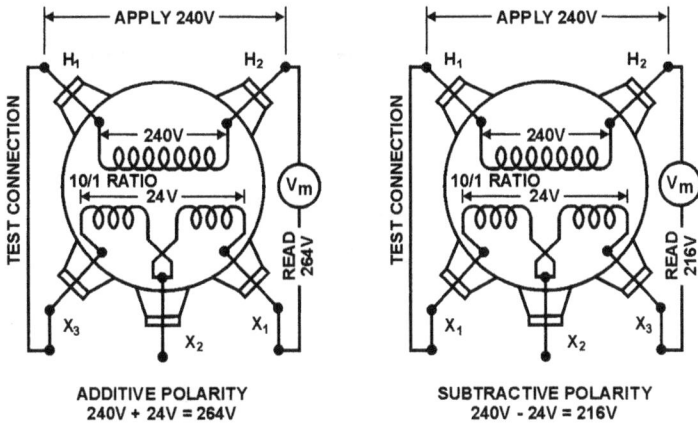

TEST CONNECTIONS FOR DETERMINING POLARITY USING ALTERNATING CURRENT
WITH REDUCED VOLTAGE FOR EXCITATION ON THE PRIMARY SIDE.
(NOTE THAT THE POSITION OF X_1 AND X_3 ARE REVERSED)

Figure 14 – Polarity Illustrated.

in figure 14. A transformer is said to have additive polarity if, when adjacent high- and low-voltage terminals are connected and a voltmeter placed across the other high- and low-voltage terminals, the voltmeter reads the sum (additive) of the high- and low-voltage windings. It is subtractive polarity if the voltmeter reads the difference (subtractive) between the voltages of the two windings. If this test is conducted, use the lowest AC voltage available to reduce potential hazards. An adjustable ac voltage source, such as a variac, is recommended to keep the test voltage low.

2.17 Single-Phase Transformer Connections for Typical Service to Buildings

Figure 15 shows a typical arrangement of bringing leads out of a single-phase distribution transformer. To provide flexibility for connection, the secondary winding is arranged in two sections.

Each section has the same number of turns and, consequently, the same voltage. Two primary leads (H_1, H_2) are brought out from the

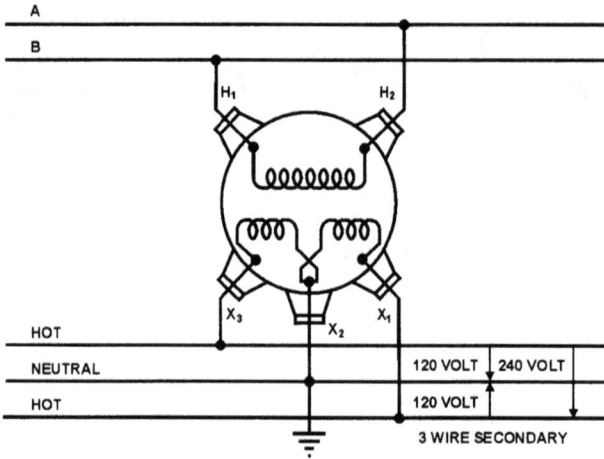

Figure 15 – Single-Phase Transformer.

top through porcelain bushings. Three secondary leads (X_1, X_2, X_3) are brought out through insulating bushings on the side of the tank, one lead from the center tap (neutral) (X_2) and one from each end of the secondary coil $(X_1$ and $X_3)$. Connections, as shown, are typical of services to homes and small businesses. This connection provides a three-wire service that permits adequate capacity at minimum cost. The neutral wire (X_2) (center tap) is grounded. A 120-volt circuit is between the neutral and each of the other leads, and a 240-volt circuit is between the two ungrounded leads.

2.18 Parallel Operation of Single-Phase Transformers for Additional Capacity

In perfect parallel operation of two or more transformers, current in each transformer would be directly proportional to the transformer capacity, and the arithmetic sum would equal one-half the total current. In practice, this is seldom achieved because of small variations in transformers. However, there are conditions for operating transformers in parallel. They are:

1. Any combination of positive and negative polarity transformers can be used. However, in all cases, numerical notations **must be followed** on both primary and secondary connections. That is H_1 connected to H_1, H_2 connected to H_2, and X_1 connected to X_1, X_2 connected to X_2, X_3 connected to X_3. Note that each subscript number on a transformer must be connected to the same subscript number on the other transformer as shown in figure 16.

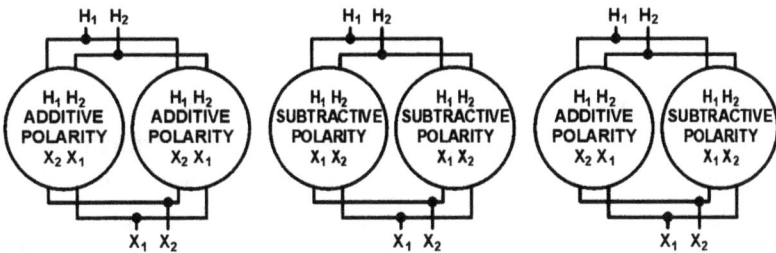

Figure 16 – Single-Phase Paralleling.

CAUTION:

With positive and negative polarity transformers, the location of X_1 and X_2 connections on the tanks will be reversed. Care must be exercised to ensure that terminals are connected, as stated above. See figure 16.

---

2. Tap settings must be identical.

3. Voltage ratings must be identical; this, of course, makes the turns ratios also identical.

4. The percent impedance of one transformer must be between $92\frac{1}{2}\%$ and $107\frac{1}{2}\%$ of the other. Otherwise, circulating currents between the two transformers would be excessive.

5. Frequencies must be identical. Standard frequency in the United States is 60 hertz and usually will not present a problem.

One will notice, from the above requirements, that paralleled transformers do not have to be the same size. However, to meet the percent impedance requirement, they must be nearly the same size. Most utilities will not parallel transformers if they are more than one standard kVA size rating different from each other; otherwise, circulating currents are excessive.

2.19 Three-Phase Transformer Connections
Three-phase power is attainable with one three-phase transformer, which is constructed with three single-phase units enclosed in the same tank or three separate single-phase transformers. The methods of connecting windings are the same, whether using the one three-phase transformer or three separate single-phase transformers.

2.20 Wye and Delta Connections
The two common methods of connecting three-phase generators, motors, and transformers are shown in figure 17. The method shown in at figure 17a is known as a delta connection, because the diagram bears a close resemblance to the Greek letter , called delta.

The other method, figure 17b, is known as the star or wye connection. The wye differs from the delta connection in that it has two phases in series. The common point "O" of the three windings is called the neutral because equal voltages exist between this point and any of the three phases.

When windings are connected wye, the voltage between any two lines will be 1.732 times the phase voltage, and the line current will be the same as the phase current. When transformers are connected delta, the line current will be 1.732 times the phase current, and the voltage between any two will be the same as that of the phase voltage.

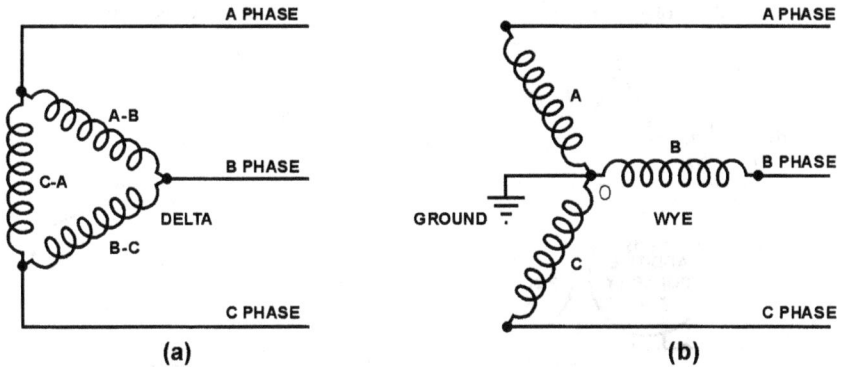

Figure 17 – Three-Phase Connections.

2.21 Three-Phase Connections Using Single-Phase Transformers

As mentioned above, single-phase transformers may be connected to obtain three-phase power. These are found at many Reclamation facilities, at shops, offices, and warehouses. The same requirements must be observed as in section 2.18, "Parallel Operation of Single-phase Transformers for Additional Capacity," with one additional requirement—in the manner connections are made between individual single-phase units. ANSI standard connections are illustrated below in the following figures. There are other angular displacements that will work but are seldom used. Do not attempt to connect single-phase units together in any combination that does keep the exact angular displacement on both primary and secondary; a dangerous short circuit could be the result. Additive and subtractive polarities can be mixed (see the following figures). These banks also may be paralleled for additional capacity if the rules are followed for three-phase paralleling discussed below. When paralleling individual three-phase units or single-phase banks to operate three phase, angular displacements must be the same.

Figure 18 shows delta-delta connections. Figure 19 shows wye-wye connections, which are seldom used at Reclamation facilities, due to

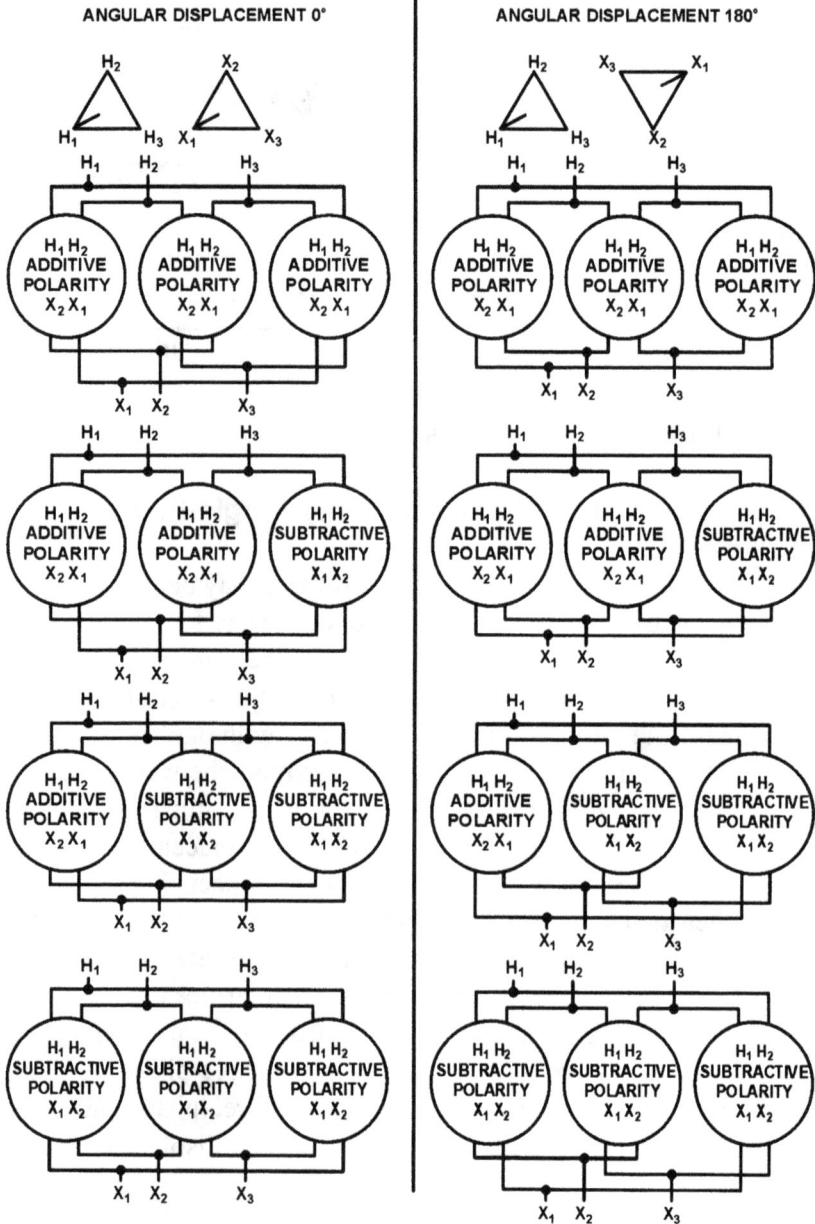

Figure 18 – Delta-Delta Connections, Single-Phase Transformers for Three-Phase Operation.

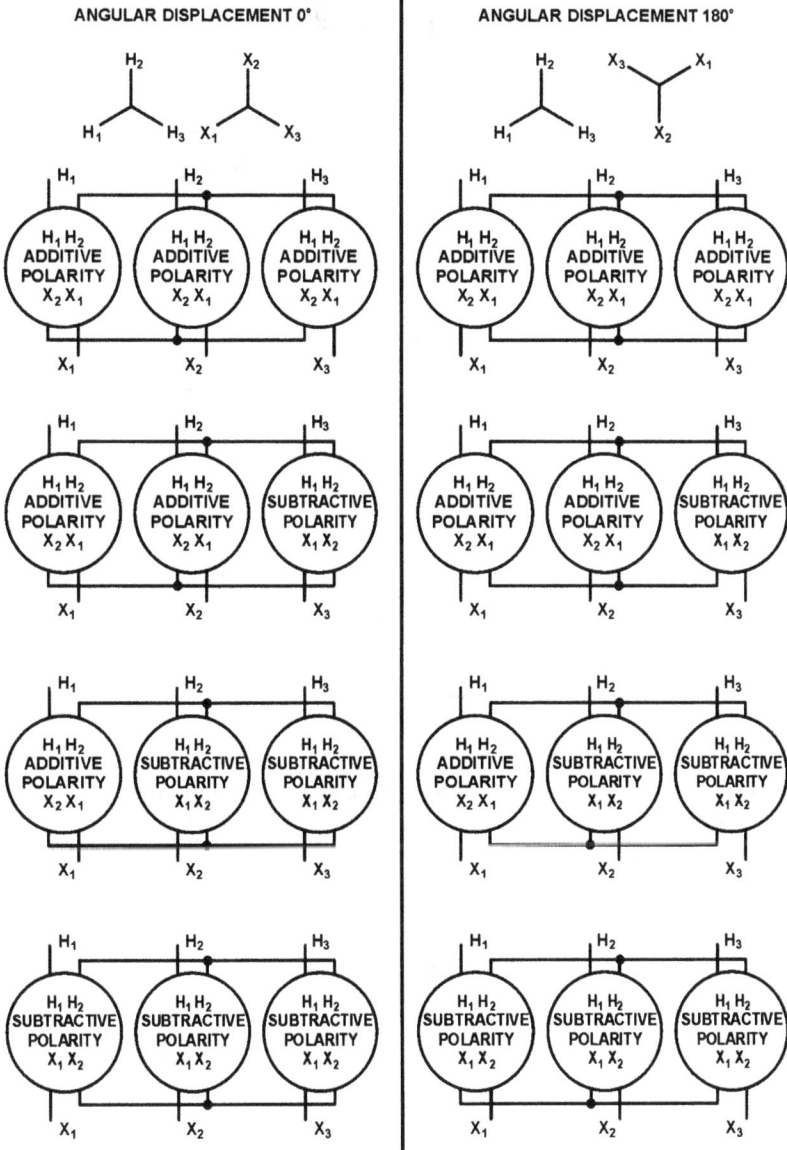

Figure 19 – Wye-Wye Connections, Using Single-Phase Transformers for Three-Phase Operation.

Note: These connections are seldom used because of 3rd harmonic problems.

inherent third harmonic problems. Methods of dealing with the third harmonic problem by grounding are listed below.

However, it is easier just to use another connection scheme (i.e., delta-delta, wye-delta, or delta-wye [figure 20]), to avoid this problem altogether. In addition, these schemes are much more familiar to Reclamation personnel.

2.22 Paralleling Three-Phase Transformers

Two or more three-phase transformers, or two or more banks made up of three single-phase units, can be connected in parallel for additional capacity. In addition to requirements listed above for single-phase transformers, phase angular displacements (phase rotation) between high and low voltages must be the same for both. The requirement for identical angular displacement must be met for paralleling any combination of three-phase units and/or any combination of banks made up of three single-phase units.

CAUTION:

This means that some possible connections will not work and will produce dangerous short circuits. See table 2 below.

For delta-delta and wye-wye connections, corresponding voltages on the high-voltage and low-voltage sides are in phase. This is known as zero phase (angular) displacement. Since the displacement is the same, these may be paralleled. For delta-wye and wye-delta connections, each low-voltage phase lags its corresponding high-voltage phase by 30 degrees. Since the lag is the same with both transformers, these may be paralleled. A delta-delta, wye-wye transformer, or bank (both with zero degrees displacement) cannot be paralleled with a delta-wye or a wye-delta that has 30 degrees of displacement. This will result in a dangerous short

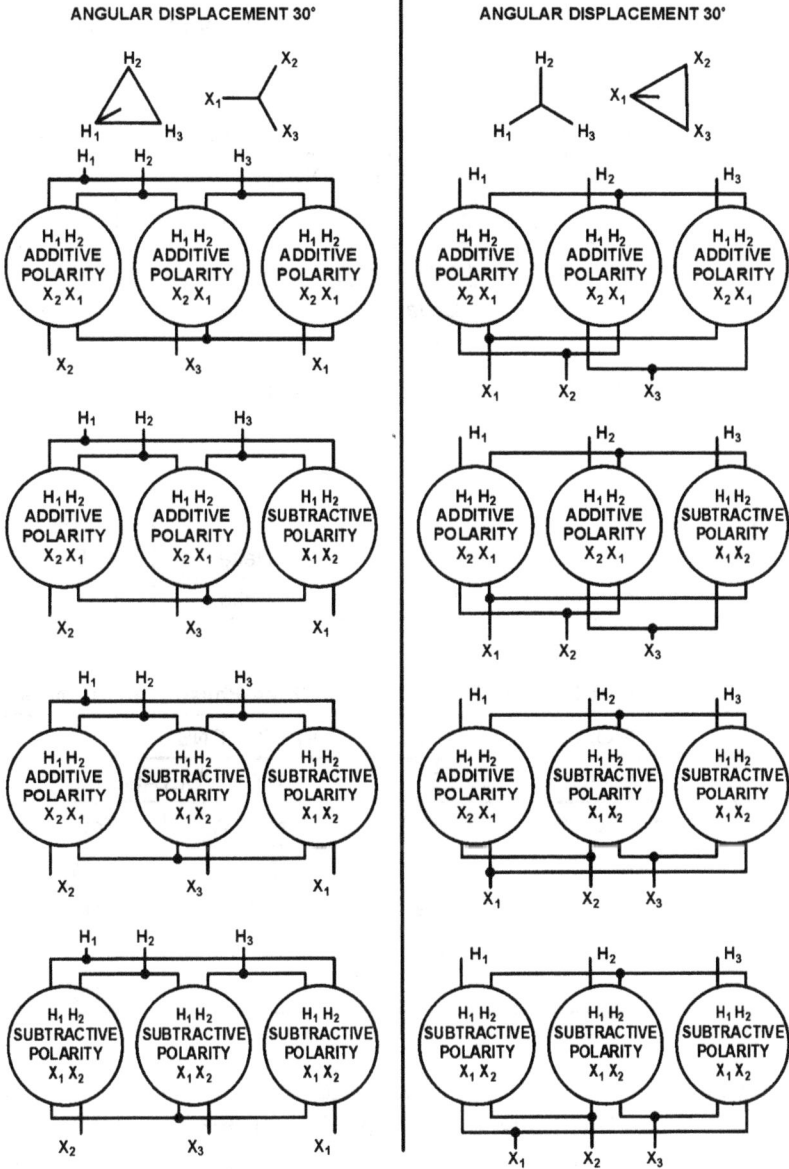

Figure 20 – Delta-Wye and Wye-Delta Connections Using Single-Phase Transformers for Three-Phase Operation.

Note: Connections on this page are the most common and should be used if possible.

circuit. Table 1 shows the combinations that will operate in parallel, and table 2 shows the combinations that will not operate in parallel.

Table 1 – Operative Parallel Connections of Three-Phase Transformers

		OPERATIVE PARALLEL CONNECTIONS		
	LOW-VOLTAGE SIDES		HIGH-VOLTAGE SIDES	
	Trans. A	Trans. B	Trans. A	Trans. B
1	Delta	Delta	Delta	Delta
2	Y	Y	Y	Y
3	Delta	Y	Delta	Y
4	Y	Delta	Y	Delta
5	Delta	Delta	Y	Y
6	Delta	Y	Y	Delta
7	Y	Y	Delta	Delta
8	Y	Delta	Delta	Y

Table 2 – Inoperative Parallel Connections of Three-Phase Transformers

		INOPERATIVE PARALLEL CONNECTIONS		
	LOW-VOLTAGE SIDE		HIGH-VOLTAGE SIDE	
	Trans. A	Trans. B	Trans. A	Trans. B
1	Delta	Delta	Delta	Y
2	Delta	Delta	Y	Delta
3	Y	Y	Delta	Y
4	Y	Y	Y	Delta

Wye-wye connected transformers are seldom, if ever, used to supply plant loads or as GSU units, due to the inherent third harmonic problems with this connection. Delta-delta, delta-wye, and wye-delta are used extensively at Reclamation facilities. Some rural electric associations use wye-wye connections that may be supplying

Reclamation structures in remote areas. There are three methods to negate the third harmonic problems found with wye-wye connections:

1. Primary and secondary neutrals can be connected together and grounded by one common grounding conductor.

2. Primary and secondary neutrals can be grounded individually using two grounding conductors.

3. The neutral of the primary can be connected back to the neutral of the sending transformer by using the transmission line neutral.

In making parallel connections of transformers, polarity markings must be followed. Regardless of whether transformers are additive or subtractive, connections of the terminals must be made according to the markings and according to the method of the connection (i.e., delta or wye).

CAUTION:

As mentioned above regarding paralleling single-phase units, when connecting additive polarity transformers to subtractive ones, connections will be in different locations from one transformer to the next.

2.23 Methods of Cooling

Increasing the cooling rate of a transformer increases its capacity. Cooling methods must not only maintain a sufficiently low average temperature but must prevent an excessive temperature rise in any portion of the transformer (i.e., it must prevent hot spots). For this reason, working parts of large transformers are usually submerged in high-grade insulating oil. This oil must be kept as free as possible from moisture and oxygen, dissolved combustible gases, and particulates.

Ducts are arranged to provide free circulation of oil through the core and coils; warmer and lighter oil rises to the top of the tank, cooler and heavier oil settles to the bottom. Several methods have been developed for removing heat that is transmitted to the transformer oil from the core and windings (figure 21).

Figure 21 – Cooling.

2.24 Oil-Filled – Self-Cooled Transformers

In small- and medium-sized transformers, cooling takes place by direct radiation from the tank to surrounding air. In oil-filled, self-cooled types, tank surfaces may be corrugated to provide a greater radiating surface. Oil in contact with the core and windings rises as it absorbs heat and flows outward and downward along tank walls, where it is cooled by radiating heat to the surrounding air. These transformers may also have external radiators attached to the tank to provide greater surface area for cooling.

2.25 Forced-Air and Forced-Oil-Cooled Transformers

Forced-air-cooled transformers have fan-cooled radiators through which the transformer oil circulates by gravity, as shown in figure 22a. Fans force air through radiators, cooling the oil.

Forced-air/oil/water-cooled transformers have a self-cooled (kVA or MVA) rating and one or more forced cooling ratings (higher kVA or MVA). Higher ratings are due to forced cooling in increasing amounts. As temperature increases, more fans or more oil pumps are turned on automatically.

FORCED COOLED TRANSFORMERS

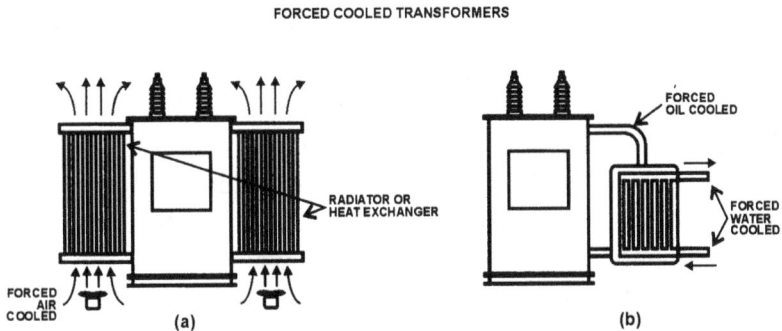

Figure 22 – Forced-Air/Oil/Water-Cooled Transformers.

The forced-cooling principle is based on a tradeoff between extra cooling and manufacturing costs. Transformers with forced-cooling have less weight and bulk than self-cooled transformers with the same ratings. In larger-sized transformers, it is more economical to add forced cooling, even though the electricity needed to operate fans and pumps increases the operating cost.

2.26 Transformer Oil

In addition to dissipating heat due to losses in a transformer, insulating oil provides a medium with high dielectric strength in which the coils and core are submerged. This allows the transformers to be more compact, which reduces costs.

Insulating oil in good condition will withstand far more voltage across connections inside the transformer tank than will air. An arc would jump across the same spacing of internal energized components at a much lower voltage if the tank had only air. In addition, oil conducts heat away from energized components much better than air.

Over time, oil degrades from normal operations, due to heat and contaminants. Oil cannot retain high dielectric strength when exposed to air or moisture. Dielectric strength declines with absorption of moisture and oxygen. These contaminants also deteriorate the paper insulation. For this reason, efforts are made to prevent insulating oil from contacting air, especially on larger power transformers. Using a tightly sealed transformer tank is impractical, due to pressure variations resulting from thermal expansion and contraction of insulating oil. Common systems of sealing oil-filled transformers are the conservator with a flexible diaphragm or bladder or a positive-pressure inert-gas (nitrogen) system. Reclamation GSU transformers are generally purchased with conservators, while smaller station service transformers have a pressurized nitrogen blanket on top of oil. Some station service transformers are dry-type, self-cooled or forced-air cooled.

2.27 Conservator System

A conservator is connected by piping to the main transformer tank that is completely filled with oil. The conservator also is filled with oil and contains an expandable bladder or diaphragm between the oil and air to prevent air from contacting the oil. Figure 23 is a schematic

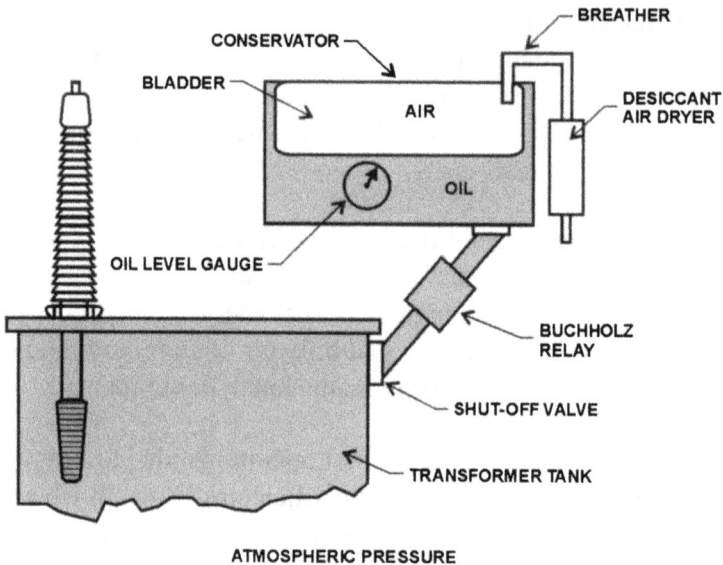

Figure 23 – Conservator with Bladder.

representation of a conservator system (figure 1 is an actual photo of a conservator). Air enters and exits the space above the bladder/ diaphragm as the oil level in the main tank goes up and down with temperature. Air typically enters and exits through a desiccant-type air dryer that must have the desiccant replaced periodically. The main parts of the system are the expansion tank, bladder or diaphragm, breather, vent valves, liquid-level gauge and alarm switch. Vent valves are used to vent air from the system when filling the unit with oil. A liquid-level gauge indicates the need for adding or removing transformer oil to maintain the proper oil level and permit flexing of the diaphragm. These are described in detail in section 4.4.

2.28 Oil-Filled, Inert-Gas System

A positive seal of the transformer oil may be provided by an inert-gas system. Here, the tank is slightly pressurized by an inert gas such as nitrogen.

The main tank gas space above the oil is provided with a pressure gauge (figure 24). Since the entire system is designed to exclude air, it must operate with a positive pressure in the gas space above the oil; otherwise, air will be admitted in the event of a leak. Smaller station service units do not have nitrogen tanks attached to automatically add gas, and it is common practice to add nitrogen yearly each fall as the tank starts to draw partial vacuum, due to cooler weather. The excess gas is expelled each summer as loads and temperatures increase.

Some systems are designed to add nitrogen automatically (figure 24) from pressurized tanks when the pressure drops below a set level. A positive pressure of approximately 0.5 to 5 pounds per square inch (psi) is maintained in the gas space above the oil to prevent ingress of air. This system includes a nitrogen gas cylinder; three-stage, pressure-reducing valve; high-and low-pressure gauges; high- and low-pressure alarm switch; an oil/condensate sump drain valve; an automatic pressure-relief valve; and necessary piping.

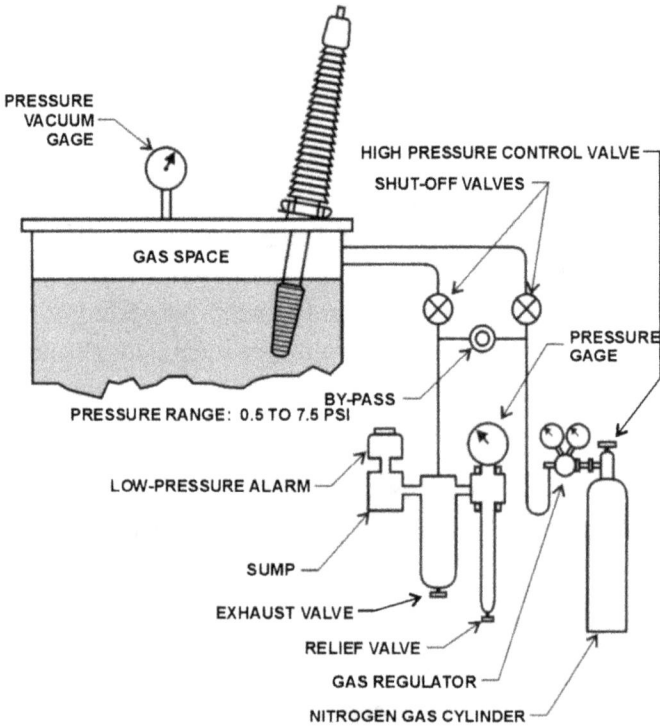

Figure 24 – Typical Transformer Nitrogen System.

The function of the three-stage, automatic pressure-reducing valves is to reduce the pressure of the nitrogen cylinder to supply the space above the oil at a maintained pressure of 0.5 to 5 psi.

The high-pressure gauge normally has a range of 0 to 4,000 psi and indicates nitrogen cylinder pressure. The low-pressure gauge normally has a range of about -5 to +10 psi and indicates nitrogen pressure above the transformer oil.

In some systems, the gauge is equipped with high- and low-pressure alarm switches to alarm when gas pressure reaches an abnormal value; the high-pressure gauge may be equipped with a pressure switch to sound an alarm when the supply cylinder pressure is running low.

A sump and drain valve provide a means for collecting and removing condensate and oil from the gas. A pressure-relief valve opens and closes to release the gas from the transformer and, thus, limit the pressure in the transformer to a safe maximum value. As temperature of a transformer rises, oil expands, and internal pressure increases, which may have to be relieved. When temperature drops, pressure drops, and nitrogen may have to be added, depending on the extent of the temperature change and pressure limits of the system. The pressurized gas system is discussed in detail in section 4.9.1.3.

2.29 Indoor Transformers

When oil-insulated transformers are located indoors, because of fire hazard, it is often necessary to isolate these transformers in a fireproof vault.

Today, dry-type transformers are used extensively for indoor installations. These transformers are cooled and insulated by air and are not encased in sealed tanks like liquid-filled units. Enclosures in which they are mounted must have sufficient space for entrance, for circulation of air, and for discharge of hot air. Dry-type transformers are enclosed in sheet metal cases with a cool air entrance at the bottom and a hot air discharge near the top. They may or may not have fans for increased air flow.

In addition to personnel hazards, indoor transformer fires are extremely expensive and detrimental to plants, requiring extensive cleanup, long outages, and lost generation. Larger indoor transformers, used for station service and generator excitation, should have differential relaying so that a fault can be interrupted quickly before a fire can ensue. Experience has shown that transformer protection by fuses alone is not adequate to prevent fires in the event of a short circuit.

3. Routine Maintenance

The following chapters address routine maintenance, as well as specific testing and diagnostic techniques and tools used to assess the condition of transformers (more detail is included for oil-filled power transformers). Some processes are often above and beyond routine maintenance work to keep the transformer operational. Transformer diagnostics require specialized equipment and training. This expertise is not expected to be maintained in every office. In some cases, it may be necessary to contact diagnostics specialists, either inside or outside Reclamation, who have the latest equipment and recent experience.

Figure 25 shows the overall transformer condition assessment methodology, linking routine maintenance and diagnostics.

3.1 Introduction to Reclamation Transformers

Standards organizations such as American National Standards Institute/Institute of Electrical and Electronic Engineers (ANSI/IEEE) consider average GSU transformer life to be 20 to 25 years. This estimate is based on continuous operation at rated load and service conditions with an average ambient temperature of 40 °C (104 degrees Fahrenheit [°F]) and a temperature rise of 65 °C. This estimate is also based on the assumption that transformers receive adequate maintenance over their service life [26]. Reclamation, Bonneville Power Administration, and Western Area Power Administration conduct regular studies to determine statistical equipment life. These studies show that average life of a Reclamation transformer is 40 years. Reclamation gets longer service than IEEE estimates because of operating at lower ambient temperatures and with lower loads. A significant number of transformers were purchased in the 1940s, 1950s, and into the 1970s. Several have been replaced, but Reclamation has many transformers that are nearing, or are already well past, their anticipated service life. We should expect transformer replacement and failures to increase, due to this age factor.

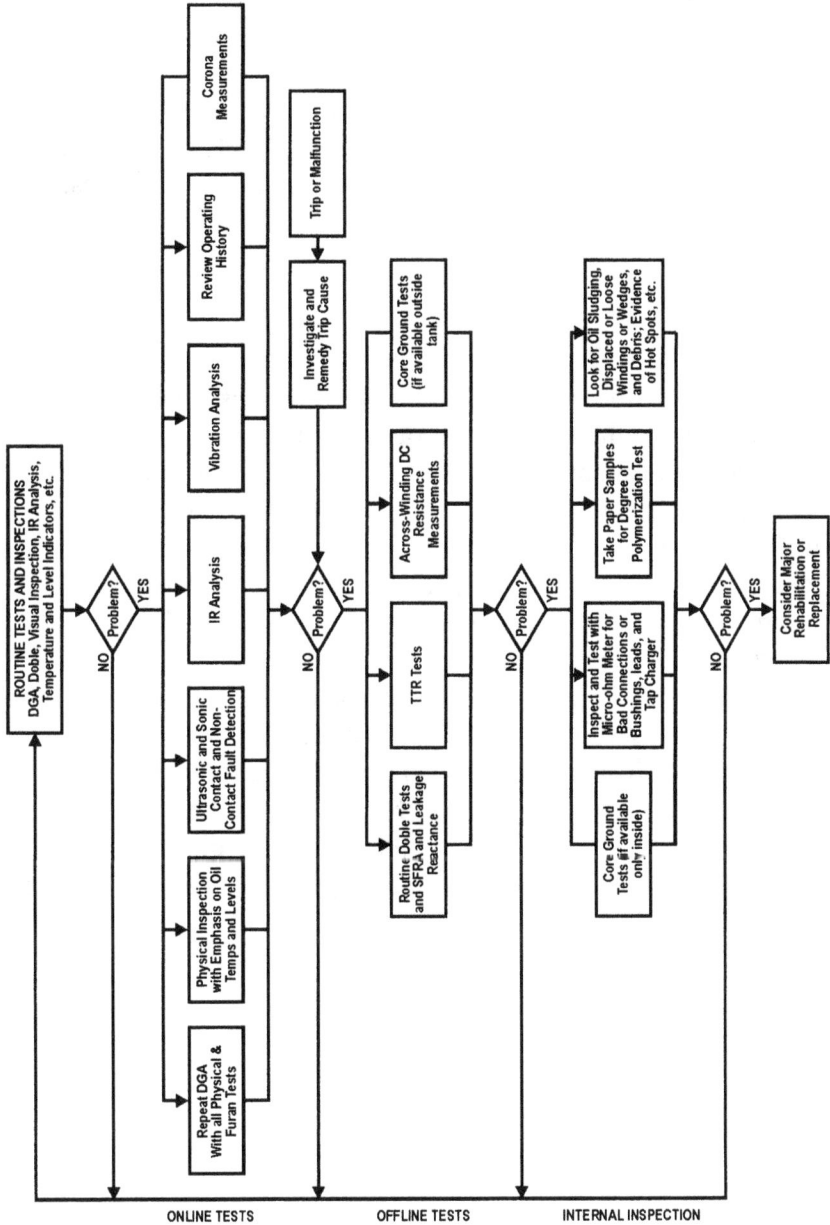

Figure 25 – Transformer Diagnostics Flowchart.

Current minimum replacement time is around 14 months; a more realistic time may be 18 months to 2 years. In the future, lead times may extend well beyond what they are today. Therefore, high-quality maintenance and accurate diagnostics are important for all transformers, but it is absolutely essential for older ones—especially for critical transformers that would cause loss of generation. It is also very important to consider providing spares for critical transformers.

3.2 Transformer Cooling Methods Introduction

Heat is one of the most common destroyers of transformers. Operation at only 8 °C above the transformer rating will cut transformer life by 50%. Heat is caused by internal losses, due to loading, high ambient temperature, and solar radiation. It is important to understand how your particular transformers are cooled and how to detect problems in the cooling systems. ANSI and IEEE require the cooling class of each transformer to appear on its nameplate. Cooling classifications, with short explanations, appear in sections 3.3 and 3.4. The letters of the class designate inside atmosphere and type or types of cooling. In some transformers, more than one class of cooling and load rating is indicated. At each step of additional cooling, the rating increases to correspond with increased cooling. Note that the letter "A" indicates air, "FA" indicates forced air (fans), "O" indicates oil, "FO" indicates forced oil (pumps), "G" indicates some type of gas, and "W" indicates there is a water/oil heat exchanger.

**DRY-TYPE TRANSFORMER MAINTENANCE SUMMARY
(See Section 3.3.)**

When new after energizing and allowing temperature and loading to stabilize	• Do an infrared scan and compare with temperature gauge, if any. • If transformer is gas filled (nitrogen [N_2]), check pressure gauge against data sheets; never allow gas pressure to fall below 1 psi. • Check loading and compare with nameplate rating. • Functionally test fans and controls for proper operation. • Functionally test temperature alarms and annunciator points. • Check area around transformer - clear of debris and parts storage. • Check transformer room for proper ventilation.
After 1 week of operation at normal loading	• Perform infrared scan and compare with temperature gauge, if any. • Check temperature gauge, if any, and compare with nameplate rating. • Check loading and compare with nameplate rating.
Annually (Note: The time between these periodic inspections may be increased if the first internal inspection of windings and connections are clean and in good condition and if loading is at or below nameplate rating.)	• Perform an infrared scan before de-energizing. • De-energize and remove panels for internal inspection. • If possible, re-energize, re-load, and do infrared (IR) inspection for hot spots and loose connections. • Use vacuum to remove as much dirt as possible. • After vacuuming, use low-pressure, dry air (20 to 25 psi) to blow off remaining dirt. Caution: Make sure air is dry. • Check for discolored copper and discolored insulation (indicates overheating). • Check tap changer and tap connections. • Check for corroded and loose connections. • Check for loose iron and damaged coils. • Check for carbon tracking on insulation and insulators. • Check for adequate electrical clearance. • Check for cracked, chipped, and loose insulators. • Check base or support insulators, including vibration isolation supports. • If windings are found dirty, add filter material to air intake ports. • Megger® high side windings to low side windings and both high and low side to ground. • Do a turns ratio test if electrical problems are found or suspected. • Check fan blades for cleanliness; remove dirt and dust. • Check fans, controls, alarms, and annunciator points. • Check pressure gauge on N_2 filled transformers; compare with weekly data sheets; never allow pressure to fall below 1 psi. • Check all bolted connections with wrenches or sockets for tightness. • Check for loose mounting for windings. • Check primary, secondary, and ground connections. • Repair all problems found in above inspections.

3.3 Dry-Type Transformers

Cooling classes of dry-type transformers are covered by ANSI/IEEE standard C57.12.01 Section 5.1 [1]. A short explanation of each class is given below.

45

1. Class AA transformers are ventilated and self-cooled. This means there are ventilation ports located in outside walls of the transformer enclosure. There are no fans to force air into and out of the enclosure, with typically no external fins or radiators. Cooler air enters the lower ports, is heated as it rises past windings, and exits the upper ventilation ports. (Although it is not repeated below; it is obvious that, in every cooling class, some heat is also removed by natural circulation of air around the outside of the enclosure.)

2. Class AFA transformers are self-cooled (A) and additionally cooled by forced circulation of air (FA). This means that there are ventilation ports for fan inlets and outlets only. (Inlets are usually filtered.) Normally, there are no additional ventilation ports for natural air circulation.

3. Class AA/FA transformers are ventilated and self-cooled (same as Class AA in item 1). In addition, they have a fan or fans providing additional forced-air cooling. Fans may be wired to start automatically when the temperature reaches a pre-set value. These transformers generally have a dual load rating— one for AA (self-cooling natural air flow) and a larger load rating for FA (forced air flow).

4. Class ANV transformers are self-cooled (A), non-ventilated (NV) units. The enclosure has no ventilation ports or fans and is not sealed to exclude migration of outside air, but there are no provisions to intentionally allow outside air to enter and exit. Cooling is by natural circulation of air around the enclosure. This transformer may have some type of fins attached outside the enclosure to increase surface area for additional cooling.

5. Class GA transformers are sealed with a gas inside (G) and are self-cooled (A). The enclosure is hermetically sealed to prevent leakage. These transformers typically have a gas, such as nitrogen or freon, to provide high dielectric and good heat removal. Cooling occurs by natural circulation of air around the outside of the enclosure. There are no fans

to circulate cooling air; however, there may be fins attached to the outside to aid in cooling.

3.3.1 Potential Problems and Remedial Actions for Dry-Type Transformer Cooling Systems [15]

It is important to keep transformer enclosures reasonably clean. It is also important to keep the area around them clear. Any items near or against the transformer impede heat transfer to cooling air around the enclosure. As dirt accumulates on cooling surfaces, it becomes more and more difficult for air around the transformer to remove heat. As a result, over time, the transformer temperature slowly rises unnoticed, reducing service life.

Transformer rooms and vaults should be ventilated. Portable fans (never water) may be used for additional cooling if necessary. A fan rated at about 100 cubic feet per minute (cfm) per kilowatt (kW) of transformer loss [6], located near the top of the room to remove hot air, will suffice. These rooms/vaults should not be used as storage.

When the transformer is new, check the fans and all controls for proper operation. After it has been energized and the loading and temperature are stable, check the temperature with an infrared (IR) camera and compare loading with the nameplate. Repeat the temperature checks after 1 week of operation.

Once each year under normal load, check transformer temperatures with an IR camera [4, 8]. If the temperature rise (above ambient) is near or above nameplate rating, check for overloading. Check the temperature alarm for proper operation. Check enclosures and vaults/rooms for dirt accumulation on transformer surfaces and debris near or against enclosures. Remove all items near enough to affect air circulation. To avoid dust clouds, a vacuum should first be used to remove excess dirt. Low-pressure (20 to 25 psi), dry compressed air may be used for cleaning after most dirt has been removed by vacuum. The transformer must be de-energized before this procedure, unless it is totally enclosed and there are no exposed energized conductors. Portable generators may be used for lighting.

After de-energizing the transformer, remove access panels and inspect windings for dirt- and heat-discolored insulation and structure problems [15]. It is important that dirt not be allowed to accumulate on windings because it impedes heat removal and reduces winding life. A vacuum should be used for the initial winding cleaning, followed by compressed air [8]. Care must be taken to ensure the compressed air is dry to avoid blowing moisture into windings. Air pressure should not be greater than 20 to 25 psi to avoid imbedding small particles into insulation. After cleaning, look for discolored copper and insulation, which indicates overheating. If discoloration is found, check for loose connections. If there are no loose connections, check the cooling paths very carefully and check for overloading after the transformer has been re-energized. Look for carbon tracking and cracked, chipped, or loose insulators. Look for and repair loose clamps, coil spacers, deteriorated barriers, and corroded or loose connections.

Check fans for proper operation, including controls, temperature switches, and alarms. Clean fan blades and filters if needed. A dirty fan blade or filter reduces cooling air flow over the windings and reduces service life. If ventilation ports do not have filters, they may be fabricated from home furnace filter material. Adding filters is only necessary if the windings are dirty upon yearly inspections.

3.4 Liquid-Immersed Transformers
Cooling classes of liquid-immersed transformers are covered by IEEE C57.12.00 Section 5.1 [2]. A short explanation of each class follows.

3.4.1 Liquid-Immersed, Air-Cooled
There are three classes in this category.

1. **Class OA:** Oil-immersed, self-cooled. Transformer windings and core are immersed in some type of oil and are self-cooled

OIL-FILLED TRANSFORMER MAINTENANCE SUMMARY

Task	After 1 Month of Service	Annually	3 to 5 Years
Before energizing, inspect and test all controls, wiring, fans, alarms, and gauges.			
In-depth inspection of transformer and cooling system;check for leaks and proper operation. Do a dissolved gas analysis (DGA).	Oil pumps load current, oil flow indicators, fans, etc. See 3.4.6 and 4. Thermometers 4.2 and 4.3. Heat exchangers. Transformer tank 4.1. Oil level gauges 4.4. Pressure relief 4.5. Do a DGA.	Oil pumps load current, oil flow indicators, fans etc. See 3.4.6 and 4. Thermometers 4.2. Heat exchangers 4.2 and 4.3. Transformer tank 4.1. Oil level gauges 4.4. Pressure relief 4.6. Do a DGA.	Check diaphragm or bladder for leaks if there is a conservator. See 4.9.
IR scan of transformer cooling system, bushings, and all wiring.	See 3.4.5.7 and 4.8.	See 3.4.5.7 and 4.8.	
Test all controls, relays, gauges; test alarms and annunciator points.	See all of section 4.	Inspect pressure relief for leaks and indication for operation (rod extension). See 4.5.	Thermometers. See 4.2 and 4.3. Oil level gauges 4.5. Inspect pressure relief 4.5. Sudden pressure relay 4.6. Buchholz relay 4.7. Test alarms, fan, and pump controls, etc. See 3.4.6.
Inspect transformer bushings.	Check with binoculars for cracks and chips; look for oil leaks and check oil levels; IR scan. See 4.8.	Check with binoculars for cracks and chips, look carefully for oil leaks and check oil levels IR scan. See 4.8.	
In-depth inspection of bushings; cleaning/waxing if needed.			Close physical inspection, cleaning/ waxing, and Doble testing, in addition to other listed inspections. See 4.8.
Doble test transformer and bushings.	Doble test transformer and bushings before energizing. See 4.8, 9.3.		See 4.8 and 9.3.
Inspect pressure controls if there is nitrogen over oil in transformer; inspect pressure gauge.	See 4.9.2.	See 4.9.2. Also see 4.9.1 to test pressure gauge if transformer has N_2 over oil with no means to automatically add N_2.	

by natural circulation of air around the outside enclosure. Fins or radiators may be attached to the enclosure to aid in cooling.

2. Class OA/FA: Liquid-immersed, self-cooled/forced air-cooled. Same as OA, with the addition of fans. Fans are usually mounted on radiators. The transformer typically has two load ratings: one with the fans off (OA) and a larger rating with fans operating (FA). Fans may be wired to start automatically at a pre-set temperature.

3. **Class OA/FA/FA:** Liquid-immersed, self-cooled/forced air-cooled/forced air-cooled. Same as OA/FA, with an additional set of fans. There typically will be three load ratings corresponding to each increment of cooling. Increased ratings are obtained by increasing cooling air over portions of the cooling surfaces. Typically, there are radiators attached to the tank to aid in cooling. The two groups of fans may be wired to start automatically at pre-set levels as temperature increases. There are no oil pumps. Oil flows through the transformer windings by the natural principle of convection (heat rising).

3.4.2 *Liquid-Immersed, Air-Cooled/Forced Liquid-Cooled*
There are two classes in this group.

1. **Class OA/FA/FOA:** Liquid-immersed, self-cooled/forced air-cooled/forced liquid, and forced air-cooled. Windings and core are immersed in some type of oil. This transformer typically has radiators attached to the enclosure. The transformer has self-cooling (OA) natural ventilation, forced air-cooling FA (fans), and forced oil-cooling (pumps) with additional forced air-cooling (FOA) (more fans). The transformer has three load ratings corresponding to each cooling step. Fans and pumps may be wired to start automatically at pre-set levels as temperature increases (figure 26).

2. **Class OA/FOA/FOA:** Liquid-immersed, self-cooled/forced oil, and forced air-cooled/forced oil, and forced air-cooled. Cooling controls are arranged to start only part of the oil

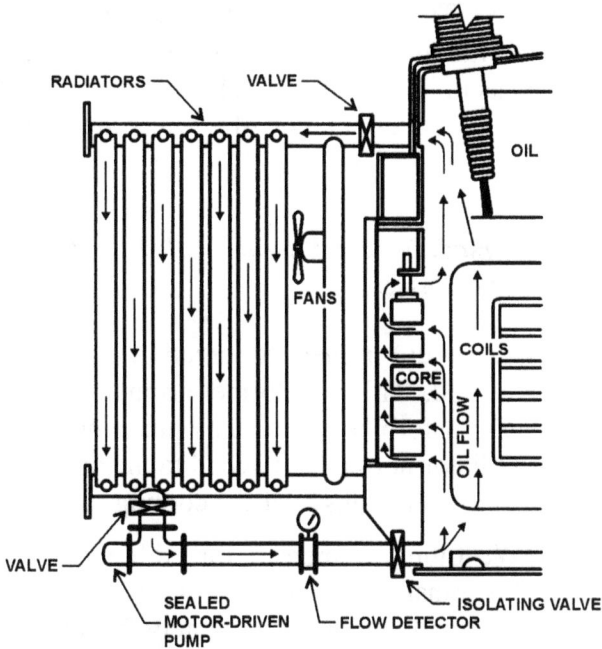

Figure 26 – Typical Oil Flow.

pumps and part of the fans for the first load rating/temperature
increase, and the remaining pumps and fans for the second load
rating increase. The nameplate will show at least three load
ratings.

3.4.3 Liquid-Immersed, Water-Cooled

This category has two classes:

1. **Class OW:** Transformer coil and core are immersed in oil.
   Typically, an oil/water heat exchanger (radiator) is attached to
   the outside of the tank. Cooling water is pumped through the
   heat exchanger, but the oil flows only by natural circulation.
   As oil is heated by the windings, it rises to the top and exits
   through piping to the radiator. As oil is cooled, it descends
   through the radiator and re-enters the transformer tank at the
   bottom.

2. **Class OW/A:** Transformer coil and core are immersed in oil. This transformer has two ratings. Cooling for one rating (OW) is obtained as in item 1. above. The self-cooled rating (A) is obtained by natural circulation of air over the tank and cooling surfaces.

3.4.4 Liquid-Immersed, Forced Liquid-Cooled
This category has two classes:

1. **Class FOA:** Liquid-immersed, forced liquid-cooled with forced air-cooled. This transformer normally has only one rating. The transformer is cooled by pumping oil (forced oil) through a radiator normally attached to the outside of the tank. Also, air is forced by fans over the cooling surface.

2. **Class FOW:** Liquid-immersed, forced liquid-cooled, water cooled. This transformer is cooled by an oil/water heat exchanger normally mounted separately from the tank. Both the transformer oil and the cooling water are pumped (forced) through the heat exchanger to accomplish cooling.

3.4.5 Potential Problems and Remedial Actions for Liquid-Filled Transformer Cooling Systems
3.4.5.1 Leaks – Tanks and radiators may develop oil leaks, especially at connections. To repair a leak in a radiator core, you must remove the radiator. Small leaks may also develop in headers or individual pipes. These small leaks possibly may be stopped by peening with a ball peen hammer. Some manufacturer's field personnel try to stop leaks by using a two-part epoxy while the transformer is under vacuum. Do not try this unless the transformer has been drained, because a vacuum may cause bubbles to form in the oil that can lodge in the winding and cause arcing. Some contractors will stop leaks by forcing epoxy into the leak area under pressure. This procedure has been successful in many cases and may be performed while the transformer is online. When all else fails, the leak may be welded with oil still in the radiator if proper precautions are

carefully observed [3, 4]. Welding with oil inside will cause gases to form in the oil. Take an oil sample for a dissolved gas analysis (DGA) before welding and 24 hours after re-energizing to identify gas increases due to welding. If the leak is bad enough, the tank may have to be drained so the leak can be repaired. Treat leaks carefully; do not ignore them. Oil leaks are serious maintenance and environmental issues and should be corrected. See *Power Equipment Bulletin 23* for further information on leak remediation.

3.4.5.2 Cleaning Radiators – Radiators may need to be cleaned in areas where deposits appear on pipes and headers. Dirt and deposits hamper heat transfer to the cooling air. Finned radiators must be cleaned with compressed air when they become dirty.

3.4.5.3 Plugged Radiators – After 1 month of service and yearly, perform an IR scan and physical inspection of radiators and transformer tanks [4, 8]. Partially plugged radiators will be cooler than those performing normally. You may also feel the radiator pipes by hand. Plugged radiator sections or individual pipes/plenums will be noticeably cooler; however, you will not be able to reach all of them. Radiators may become plugged with sludge or foreign debris; this usually occurs in water tubes on the oil/water heat exchanger. Do not forget to check the bleed line for two-walled heat exchangers.

If plugged radiators are discovered, they must be corrected as soon as possible. Some radiators are attached to the main tank with flanges and have isolating valves. These may be removed for cleaning and/or leak repair without draining oil from the transformer. If radiators are attached directly to the main tank without isolating valves, oil must be drained before cleaning them. If radiators are plugged with sludge, the transformer probably is sludged up also. In this case, the oil should be reprocessed and the transformer cleaned internally. Competent contractors should be obtained if this is necessary.

3.4.5.4 Sludge Formation – If temperature seems to be slowly increasing while the transformer is operating under the same load, check the DGA for moisture, oxygen, acid number, and the interfacial tension (IFT). The combination of oxygen and moisture causes sludging, which may be revealed by a low IFT number and/or acid number. Sludge will slowly build up on windings and core, and the temperature will increase over time.

3.4.5.5 Valve Problems – If your transformer has isolating valves for radiators, check to make sure they are fully open on both top and bottom of the radiators. A broken valve stem may cause the valve to be fully or partially closed, but it will appear that the valve is open.

3.4.5.6 Mineral Deposits – Don't even think about spraying water on the radiators or tank to increase cooling, except in the most dire emergency. Minerals in the water will deposit on radiators as water evaporates and are almost impossible to remove. These minerals will reduce the efficiency of cooling still further. Using additional fans to blow air on radiators and/or the transformer tank is a better alternative [4].

3.4.5.7 Low Oil Level – One IR scan performed on a transformer running at higher than normal temperature revealed that the oil level was below the upper radiator inlet pipe, which prevented oil circulation. The oil level indicator was defective and stuck on normal. These indicators must be tested, as mentioned below.

3.4.6 Cooling System Inspections

After 1 month of service and yearly, inspect and test the fans. Look at the fans whenever you are around transformers in the switchyard or in the powerplant. If it is a hot day and transformers are loaded, all the fans should be running. If one fan is stopped and the rest of the group is running, the inactive fan should be repaired. During an inspection, the temperature controller should be adjusted to start all the fans. Listen for unusual noises from fan bearings and loose blades and repair or replace faulty fans. Bad bearings can also be detected with an IR scan if the fans are running.

After 1 month of service and yearly, inspect and test the oil pumps. Inspect piping and connections for leaks. Override the temperature controller so that the pump starts. Check the oil pump motor current on all three phases with an accurate ammeter; this will give an indication if oil flow is correct and if unusual wear is causing additional motor loading. Record this information for later comparison, especially if there is no oil flow indicator. If the motor load current is low, something is causing low oil flow. Carefully inspect all valves to make sure they are fully open. A valve stem may break and leave the valve partially or fully closed, even though the valve handle indicates the valve is fully open. Pump impellers have been found loose on the shaft, reducing oil flow. Sludge buildup or debris in lines can also cause low oil flow. If motor load current is high, this may indicate impeded pump rotation. Listen for unusual noises. Thrust bearing wear results in the impeller advancing on the housing. An impeller touching the housing makes a rubbing sound, which is different from the sound of a failing motor bearing. If this is heard, remove the pump motor from the housing and check impeller clearance. Replace the thrust bearing if needed, and replace the motor bearings if the shaft has too much play or if noise is unusual.

Three-phase pumps will run and pump some oil even if they are running backwards. Vane type oil-flow meters will indicate flow on this low amount. The best indication of pump running backwards is that sometimes the pump will be very noisy. The motor load current may also be lower than for full load. If this is suspected, due to the extra noise and higher transformer temperature, the pump should be checked for proper rotation. Reverse two phase leads if this is encountered. [4]

After 1 month of service and yearly, check the oil flow indicator. It has a small paddle which extends into the oil stream and may be either on the suction or discharge side of the pump. A very low flow of only about 5-feet-per-second velocity causes the flag to rotate, which indicates normal flow. Flow can be too low, and the indicator will still show flow. If there is no flow, a spring returns the flag to the off position, and a switch provides an alarm. With control power on the switch, open the pump circuit at the motor starter and make sure the correct alarm point activates when the pump stops. Check that the pointer is in the right position when the pump is off and when it is running. Pointers can stick and fail to provide an alarm when needed. Oil flow may also be checked with an ultrasonic flowmeter. Ultrasonic listening devices can detect worn bearings, rubbing impellers, and other unusual noises from oil pumps.

Pumps can pull air in through gaskets on the suction side of the pumps. The suction (vacuum) on the intake side of the pump can pull air through gaskets that are not tight. Pump suction has also been known to pull air through packing around valve stems in the suction side piping. This can result in dangerous bubbles in the transformer oil and may cause the gas detector or Buchholz relay to operate. Dissolved gas analysis will show a big increase in oxygen and nitrogen content [4]. High oxygen and nitrogen content can also be caused by gasket leaks elsewhere.

After 1 month of service and yearly, inspect water-oil heat exchangers. Test and inspect the pumps, as mentioned above. Look for and repair leaks in piping and heat exchanger body. Examine the latest dissolved gas analysis results for dissolved moisture and free water. If free water is present and there are no gasket leaks, the water portion of the water-oil heat exchanger must be pressure tested. A leak may have developed, allowing water to migrate into the transformer oil, which can destroy the transformer. If the heat exchanger piping is double walled, check the drain for water or oil; check manufacturer's instruction manual.

4. Oil-Filled Transformer Inspections

A transformer maintenance program must be based on thorough routine inspections. These inspections must take place in addition to normal daily/weekly data gathering trips to check oil levels and temperatures. Some monitoring may be done remotely using supervisory control and data acquisition (SCADA) systems, but this can never substitute for thorough inspections by competent maintenance or operations people.

After 1 month of service, and once each year, make an indepth inspection of oil-filled transformers. Before beginning, carefully inspect the temperature and oil level data sheets. If temperature, pressure, or oil level gauges never change, even with seasonal temperature and loading changes, something is wrong. The gauge may be stuck, or data sheets may have been filled in incorrectly. Examine the DGAs for evidence of leaks or other problems.

4.1 Transformer Tank
Check for excessive corrosion and oil leaks. Pay special attention to flanges and gaskets (bushings, valves, and radiators) and the lower section of the main tank. Report oil leaks to maintenance, and pay special attention to the oil level indicator if leaks are found. Severely corroded spots should be wire brushed and painted with a rust inhibitor.

4.2 Top Oil Thermometers
These thermometers are typically sealed, spiral-bourdon-tube dial indicators with liquid-filled bulb sensors. The bulb is normally inside a thermometer well, which penetrates the tank wall into the oil near the top of the tank. As oil temperature increases in the bulb, liquid expands, which expands the spiral tube. The tube is attached to a pointer that indicates temperature. These pointers may also have electrical contacts to trigger alarms and start cooling fans as temperature increases. An extra pointer, normally red, indicates maximum temperature since the last time the indicator was reset. This red pointer rises with the main pointer but will not decrease unless manually reset; thus, it always indicates the highest

temperature reached since it was reset. See the instruction manual on your specific transformer for details.

4.3 Winding Temperature Thermometers

These devices are supposed to indicate the hottest spot in the winding, based on the manufacturer's heat run tests. At best, this device is only accurate at top nameplate rated load and only if it is not out of calibration [18]. They are not what their name implies and can be misleading. They are only winding hottest-spot simulators, which are not very accurate. Normally, there is no temperature sensor embedded in the winding hot spot. At best, they provide only a rough approximation of hot spot winding temperature and should not be relied on for accuracy. They can be used to turn on additional cooling or activate alarms as the top oil thermometers do.

Winding temperature thermometers work the same way as the top oil thermometer (discussed in section 4.2), except that the bulb is in a separate thermometer well near the top of the tank. A wire-type heater coil is either inserted into, or wrapped around, the thermometer well, which surrounds the temperature sensitive bulb. In some transformers, a current transformer (CT) is around one of the three winding leads and provides current directly to the heater coil in proportion to winding current. In other transformers, the CT supplies current to an auto-transformer that supplies current to the heater coil. The heater warms the bulb, and the dial indicates a temperature, but it is not the true hottest-spot temperature.

These devices are calibrated at the factory by changing taps on either the CT or the autotransformer, or by adjusting the calibration resistors in the control cabinet. These devices normally cannot be field calibrated or tested, other than testing the thermometer, as mentioned. The calibration resistors can be adjusted in the field if the manufacturer provides calibration curves for the transformer. In practice, most winding temperature indicators are out of calibration, and their readings are meaningless. These temperature indications should not be relied upon for loading operations or maintenance decisions.

Fiber optic temperature sensors can be imbedded directly into the winding as the transformer is being built; these sensors are much more accurate. This system is available as an option on new transformers at an increased cost, which may be worthwhile, since the true winding "hottest-spot" temperature is critical when higher loading is required.

Thermometers can be removed without lowering the transformer oil if they are in a thermometer well. Check your transformer instruction manual. Look carefully at the capillary tubing between the thermometer well and the dial indicator. If the tubing has been pinched or accidentally struck, it may be restricted. This is not an obvious defect, but it can cause the dial pointer to lock in one position. If this defect is found, the whole gauge must be returned to the factory for repair or replacement; it cannot be repaired in the field. Look for a leak in the tubing system; the gauge reading will be very low and must be replaced if a leak is discovered.

Every 3 to 5 years, and if trouble is suspected, test the thermometer. Suspend the thermometer's indicator bulb and an accurate mercury thermometer in an oil bath. Do not allow either thermometer to touch the side or bottom of the container. Heat the oil on a hotplate, while stirring, and compare the two thermometers while the temperature increases. If a magnetic stirring/heating plate is available, it is more effective than hand stirring. Pay particular attention to the upper temperature range at which your transformers normally operate (50 °C to 80 °C). An ohmmeter should also be used to check switch operations. If either dial indicator is more than 5 °C different than the mercury thermometer, it should be replaced with a spare. A number of spares should be kept, based on the quantity of transformers at the plant. Oil bath test kits are available from the Qualitrol Company. After calling for Qualitrol authorization

(716-586-1515), you can ship defective dial thermometers for repair and calibration to: Qualitrol Company, 1387 Fairport Road, Fairport, New York 14450.

The alarms and other functions should also be tested to see if the correct annunciator points activate, pumps/fans operate, etc.

If the temperature gauge cannot be replaced or sent to the factory for repair, place a temperature correction factor on your data form to add to the dial reading, so that the correct temperature will be recorded. Also, lower the alarm and pump-turn-on settings by this same correction factor. Since these are pressure-filled systems, the indicator will typically read low if it is out of calibration. Field testing has shown some of these gauges reading 15 °C to 20 °C lower than actual temperature. This is hazardous for transformers because it will allow them to continuously run hotter than intended, due to delayed alarms and cooling activation. If thermometers are not tested and errors corrected, transformer service life may be shortened or premature failure may occur.

4.3.1 Temperature Indicators Online

Check all temperature indicators while the transformer is online. The winding temperature indicator should read approximately 15 degrees above the top oil temperature. If this is not the case, one or both temperature indicators are malfunctioning. Check the top oil temperature next to the top oil indicator's thermowell with an infrared camera. Compare the readings with the top oil indicator. Reset all maximum indicator hands on the temperature indicating devices after recording the old maximum temperature readings. High temperature may mean overloading, cooling problems, or problems with windings, core, or connections.

4.3.2 Temperature Indicators Offline

When the transformer is offline and has cooled to ambient temperature, check the top oil and winding temperature indicators; both should read the same. If not, one or both temperature indicators are mal-functioning. Check the calibration according to the proper procedure. Also, compare these readings with the indicated temperature on the conservator oil level indicator; all three should agree.

4.4 Oil Level Indicators

After 1 month of service, and every 3 to 5 years, check the tank oil level indicators. These are float operated, with the float mechanism magnetically coupled through the tank wall to the dial indicator. As level increases, the float rotates a magnet inside the tank. Outside the tank, another magnet follows (rotates), which moves the pointer. The center of the dial is normally marked with a temperature of 25 °C (77 °F). High and low level points are also marked to follow level changes as the oil expands and contracts with temperature changes. The proper way to determine accurate oil level is to first look at the top oil temperature indicator. After determining the temperature, look at the level gauge. The pointer should be at a reasonable level corresponding to the top oil temperature. Calibrate or replace the conservator oil level indicator if needed, but only after checking the top oil temperature indicator as shown in the above section. If the transformer is fully loaded, the top oil temperature will be high, and the level indicator should be near the high mark. If the transformer is de-energized and the top oil temperature is near 25 °C, the oil level pointer should be at or near 25 °C. See figure 27. Reference also IEEE 62-1995™ [20], section 6.6.2.

To check the level indicator, remove the outside mechanism for testing without lowering transformer oil. After removing the gauge, hold a magnet on the back of the dial and rotate the magnet; the dial indicator should also rotate. If it fails to respond or if it drags or sticks, replace it. As mentioned above, defective units can be sent to the factory for repair.

There may also be electrical switches for alarms and, possibly, for tripping off the transformer when the tank level falls. These switches

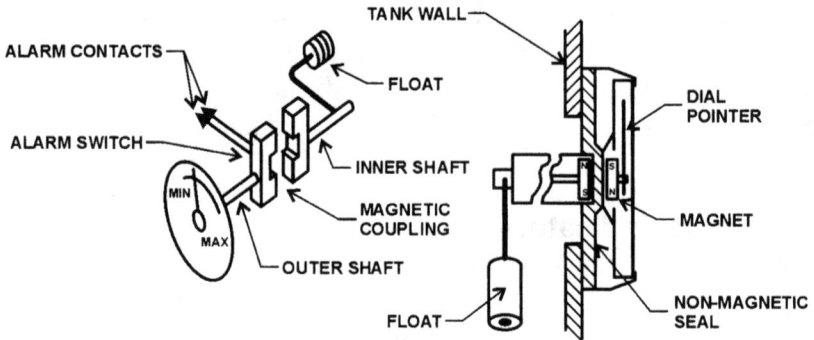

Figure 27 – Oil Level Indicator.

should be checked with an ohmmeter for proper operation. The alarm/tripping circuits should also be tested to see if the correct annunciator points and relays respond. See the transformer instruction book for information on your specific indicator.

Figure 28 – Conservator Oil Level.

If oil has had to be lowered in the transformer or conservator for other reasons (e.g., inspections), check the oil level float mechanism (figure 28). Rotate the float mechanism by hand to check for free movement.

Check the float visually to make sure it is secure to the arm and to ensure that the arm is in the proper shape. Some arms are formed (not straight).

4.5 Pressure Relief Devices

These devices are the transformers' last line of defense against excessive internal pressure. In the event of a fault or short circuit, the resultant arc instantly vaporizes surrounding oil, causing a rapid buildup of gaseous pressure. **If the pressure relief device does not operate properly and pressure is not sufficiently relieved within a**

few milliseconds, a catastrophic tank rupture can result, spreading flaming oil over a wide area. Two types of pressure relief devices are discussed below. Consult your transformer's instruction manual for specifics.

CAUTION:

Never paint pressure relief devices because paint can cause the plunger or rotating shaft to stick. Then, the device might not relieve pressure, which could lead to catastrophic tank failure during a fault. Look at the top of the device; on newer units, a yellow or blue button should be visible. If these have been painted, the button will be the same color as the tank. On older units, a red flag should be visible; if it has been painted, it will be the same color as the tank.

If the pressure relief devices have been painted, they should be replaced. It is virtually impossible to remove all paint from the mechanism and be certain the device will work when needed.

4.5.1 Newer Pressure Relief Devices

Newer pressure relief devices are spring-loaded valves that automatically re-close following a pressure release. The springs are held in compression by the cover and press on a disc which seals an opening in the tank top. If pressure in the tank exceeds operating pressure, the disc moves upward and relieves pressure. As pressure decreases, the springs re-close the valve. After operating, this device leaves a brightly colored rod (bright yellow for oil, blue for silicone) exposed approximately 2 inches above the top. This rod is easily seen upon inspection, although it is not always visible from floor level. The rod may be reset by pressing on the top until it is again recessed into the device. The switch must also be manually reset. A relief device is shown in the open position in figure 29. Figure 30 also shows a pressure relief device with the yellow indicating arm.

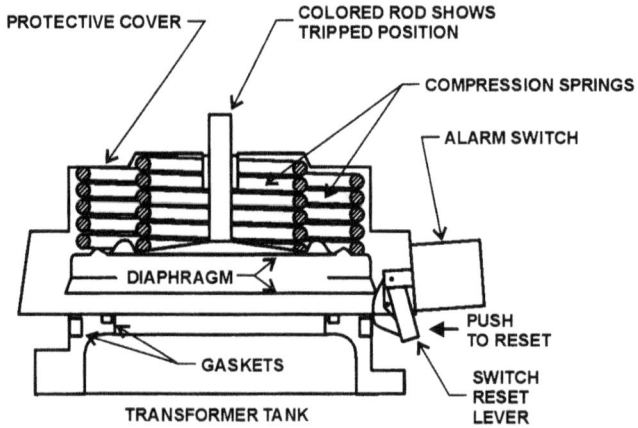

Figure 29 – Pressure Relief Device.

Figure 30 – Photograph of a Pressure
Relief Device.

CAUTION:

Do not re-energize a transformer after the pressure relief
device has operated and relays have de-energized the
transformer, until extensive testing to determine and correct
the cause has been completed. Explosive, catastrophic
failure could be the result of energization after this device
has operated.

CAUTION:

Bolts that hold the device to the tank may be loosened
safely, but never loosen screws that hold the cover to the
flange without referring to the instruction manual and using
great care. Springs that oppose tank pressure are held in
compression by these screws, and their stored energy could
be hazardous.

Once each year, and as soon as possible after a known through-fault or
internal fault, inspect pressure devices to see if they have operated.
This must be done from a high-lift bucket if the transformer is
energized. Look at each pressure relief device to see if the yellow
(or blue) button is visible. If the device has operated, about 2 inches
of the colored rod will be visible. Each year, test the alarm circuits by
operating the switch by hand and making sure the correct annunciator
point is activated. If the relief device operates during operation, do not
re-energize the transformer; Doble and other testing may be required
before re-energizing, and an oil sample should be sent for analysis.

Every 3 to 5 years, when doing other maintenance or testing, if the
transformer has a conservator, examine the top of the transformer tank
around the pressure relief device. If oil is visible, the device is
leaking, either around the tank gasket or relief diaphragm. If the
device is 30 years old, replace the whole unit. A nitrogen blanketed
transformer will use a lot more nitrogen if the relief device is leaking;
they should be tested as described below.

A test stand with a pressure gauge may be fabricated to test the
pressure relief function. Current cost of a pressure relief device
is about $600, so testing instead of replacement may be prudent.
Have a spare pressure relief device on hand so that the tank will
not have to be left open. If the tank top or pressure relief device
has gasket limiting grooves, always use a nitrile replacement gasket;
if there are no grooves, use a cork-nitrile gasket. Although relief
devices themselves do not leak often, the gasket may leak.

4.5.2 Older Pressure Relief Devices

Older pressure relief devices have a diaphragm and a relief pin that is destroyed each time the device operates and must be replaced.

CAUTION:

Replacement parts of an older pressure relief device must be replaced with exact duplicate parts; otherwise, the operating relief pressure of the device will be wrong.

The relief pin determines operating pressure; a number, which is the operating pressure, normally appears on top of the pin. Check your specific transformer instruction manual for proper catalog numbers. Do not assume you have the right parts or that correct parts have been previously installed—look it up. If the operating pressure is too high, a catastrophic tank failure could result.

On older units, a shaft rotates, operates alarm/trip switches, and raises a small red flag when the unit releases pressure. If units have been painted or are more than 30 years old, they should be replaced with the new model as soon as it is possible to have a transformer outage.

Once each year, and as soon as possible after a through-fault or internal fault, examine the indicator flag to see if the device has operated. The flags must be examined from a high-lift bucket if the transformer is energized. A clearance must be obtained to test, repair, or reset the device. See the instruction manual for your specific transformer. Test alarm/trip circuits by operating the switch by hand. Check to make sure the correct annunciator point activates.

Every 3 to 5 years, when doing other maintenance or testing, examine the top of the transformer tank around the pressure relief device. If the transformer has a conservator and oil is visible, the device is leaking, either around the tank gasket or relief diaphragm. The gasket and/or device must be replaced. Before ordering, make sure that the new device will fit the same tank opening. Most devices are made by the Qualitrol Company; contact the manufacturer to obtain a correct replacement.

4.6 Sudden Pressure Relay

Internal arcing in an oil-filled power transformer can instantly vaporize surrounding oil, generating gas pressures that can cause catastrophic failure, rupture the tank, and spread flaming oil over a large area. This can damage or destroy other equipment, in addition to the transformer, and present extreme hazards to workers.

The relay is designed to detect a sudden pressure increase caused by arcing. This relay is very sensitive and will operate if the pressure rises even slightly. If a very small pressure change occurs caused by a small electrical fault inside the tank, this relay will alarm. The relay is set to operate before the pressure relief device. The control circuit should de-energize the transformer and provide an alarm. The relay will ignore normal pressure changes such as oil-pump surges, temperature changes, etc.

Modern sudden pressure relays consist of three bellows (see figure 31) with silicone sealed inside. Changes in pressure in the transformer deflect the main sensing bellows. Silicone inside acts on two control bellows arranged like a balance beam (one on each side). One bellows senses pressure changes through a small orifice. The opening is automatically changed by a bimetallic strip to adjust for normal temperature changes of the oil. The orifice delays pressure changes in this bellows. The other bellows responds to immediate pressure changes and is affected much more quickly. Pressure difference tilts the balance beam and activates the switch. This type of relay automatically resets when the two bellows again reach pressure equilibrium. If this relay operates, do not re-energize the transformer until you have determined the exact cause and corrected the problem.

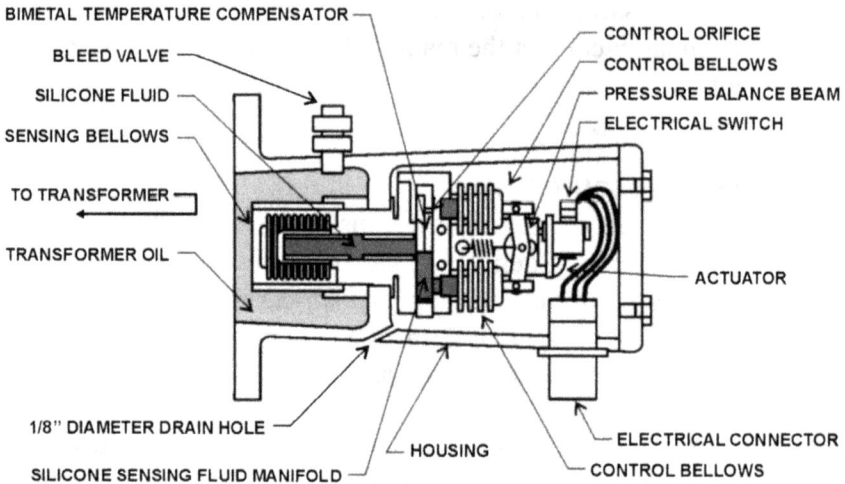

Figure 31 – Sudden Pressure Relay, Section.

Old style sudden pressure relays have only one bellows. A sudden excessive pressure within the transformer tank exerts pressure directly on the bellows, which moves a spring-loaded operating pin. The pin operates a switch that provides alarm and breaker trip. After the relay has operated, the cap must be removed and the switch must be reset to normal by depressing the reset button.

Once every 3 to 5 years, the sudden pressure relay should be tested according to manufacturer's instructions. Generally, only a squeeze-bulb and pressure gauge (5 psi) are required. Disconnect the tripping circuit and use an ohmmeter to test for relay operation. Test the alarm circuit and verify that the correct alarm point is activated. Use an ohmmeter to verify the trip signal is activated or, if possible, apply only control voltage to the breaker and make sure the tripping function operates. Consult the manufacturer's manual for your specific transformer's detailed instructions.

4.6.1 Testing Suggestion

Figure 32 shows an example relay. Inspect the isolation valve to ensure it is open. With the transformer offline and under clearance, functionally test the sudden pressure relay by slowly closing the isolating valve. Leave it closed for a few seconds and reopen the valve very suddenly; this should activate the alarm. If the alarm does not activate, test the relay. If the relay fails to function, replace it with a new one.

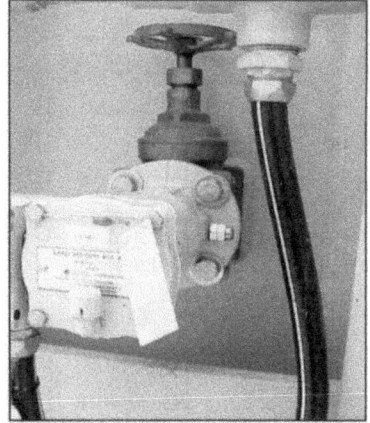

Figure 32 – Photograph of a Sudden Pressure Relay.

4.7 Buchholz Relay (Found Only on Transformers with Conservators)

The Buchholz relay (figures 33 and 34) has two oil-filled chambers with floats and relays arranged vertically—one over the other. If high eddy currents, local overheating, or partial discharges occur within the tank, bubbles of resultant gas rise to the top of the tank. These bubbles rise through the pipe between the tank and the conservator. As gas bubbles migrate along the pipe, they enter the Buchholz relay and rise into the top chamber. As gas builds up inside the chamber, it displaces the oil, which decreases the oil level. The top float descends with the oil level until it passes a magnetic switch, which activates an alarm. The bottom float and relay cannot be activated by additional gas buildup. The float is located slightly below the top of the pipe so that once the top chamber is filled, additional gas goes into the pipe and continues up to the conservator. Typically, inspection windows are provided so that the amount of gas and relay operation may be viewed during testing. If the oil level falls low enough (conservator empty), the bottom float activates the switch contacts in the bottom chamber.

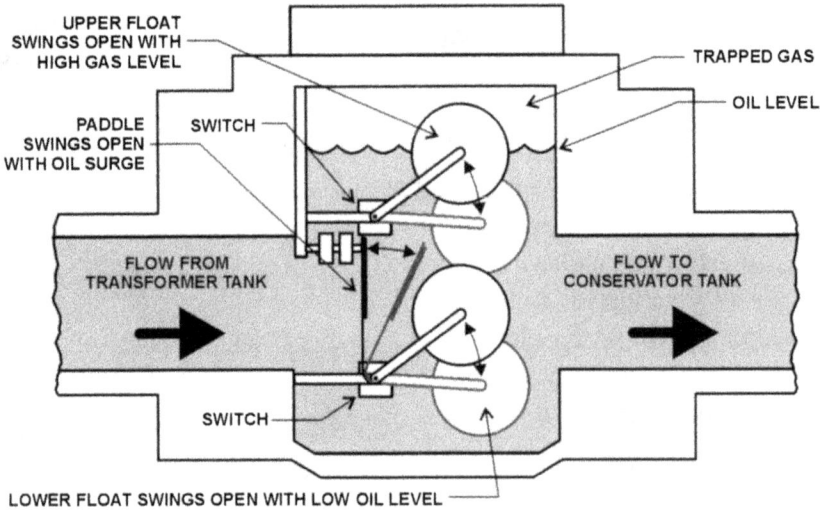

Figure 33 – Buchholz Relay, Section.

Figure 34 – Photograph of a Buchholz Relay.

These contacts are typically connected to cause the transformer to trip. This relay also serves a third function, similar to the sudden pressure relay. A magnetically held paddle attached to the bottom float is positioned in the oil-flow stream between the conservator and transformer tank. Normal flows resulting from temperature changes are small and bypass below the paddle. If a fault occurs in the transformer, a pressure wave (surge) is created in the oil. This surge travels through the pipe and displaces the paddle. The paddle activates the same magnetic switch as the bottom float mentioned above, which trips the transformer. The flow rate at which the paddle activates the relay is normally adjustable. See your specific transformer's instruction manual for details.

Once every 3 to 5 years, while the transformer is de-energized, functionally test the Buchhholz relay by pumping a small amount of air into the top chamber with a squeeze-bulb hand pump. Watch the float operation through the window (center in figure 32). Check to make sure the correct alarm point has been activated. Open the bleed valve and vent air from the chamber. The bottom float and switching cannot be tested with air pressure. On some relays, a rod is provided so that you can test both bottom and top sections by pushing the floats down until the trip points are activated. If possible, verify that the breaker will trip with this operation. A volt-ohmmeter may also be used to check the switches. If these contacts activate during operation, it means that the oil level is very low, a pressure wave has activated (bottom contacts), or the transformer is gassing (top contacts). If this relay operates, do not re-energize the transformer until you have determined the exact cause.

If a small amount of gas is found in this relay when the transformer is new (a few months after startup), it is probably just air that has been trapped in the transformer structure and is now escaping; there is little cause for concern.

If the transformer has been online for some time (service aged), and gas is found in the Buchholz, oil samples must be sent to the lab for DGA and extensive testing. Consult with the manufacturer and other transformer experts. A definite cause of the gas bubbles must be determined and corrected before re-energization of the transformer.

4.8 Transformer Bushings: Testing and Maintenance of High-Voltage Bushings

When bushings are new, they should be Doble tested as an acceptance test. Refer to the M4000 Doble test set instructions, the *Doble Bushing Field Test Guide* [9], and the manufacturer's data for guidance on acceptable results.

CAUTION:

Do not test a bushing while it is in its wood shipping crate, or while it is lying on wood. Wood does not insulate as well as porcelain and will cause the readings to be inaccurate. Keep the test results as a baseline record to compare with future tests.

After 1 month of service and yearly, check the external porcelain for cracks and/or contamination (requires binoculars). There is no "perfect insulator"; a small amount of leakage current always exists. This current "leaks" through and along the bushing surface from the high-voltage conductor to ground. If the bushing is damaged or heavily contaminated, leakage current becomes excessive, and visible evidence may appear as carbon tracking (treeing) on the bushing surface. Flashovers may occur if the bushings are not cleaned periodically.

Look carefully for oil leaks. Check the bushing oil level by viewing the oil-sight glass or the oil level gauge. When the bushing has a gauge with a pointer, look carefully, because the oil level should vary slightly with temperature changes. If the pointer never changes, even with wide ambient temperature and load changes, the gauge should be checked at the next outage. A stuck gauge pointer coupled with a small oil leak can cause explosive failure of a bushing, damaging the transformer and other switchyard equipment. A costly extended outage is the result.

If the oil level is low and there is an external oil leak, check the bolts for proper torque and the gasket for proper compression. If torque and compression are correct, the bushing must be replaced with a spare. Follow instructions in the transformer manual carefully. It is very important that the correct type of gasket be installed and the correct compression be applied. A leaky gasket is probably also leaking water and air into the transformer, so check the most recent transformer DGA for high moisture and oxygen.

If the oil level is low and there is no visible external leak, there may be an internal leak around the lower seal into the transformer tank. If possible, re-fill the bushing with the same oil and carefully monitor the level and the volume it takes to fill the bushing to the proper level. If it takes more than 1 quart of oil, make plans to replace the bushing. The bushing must be sent to the factory for repair, or it must be junked; it cannot be repaired in the field.

CAUTION:

Never open the fill plug of any bushing if it is at an elevated temperature. Some bushings have a nitrogen blanket on top of the oil, which pressurizes as the oil expands. Always consult the manufacturer's instruction manual which will give the temperature range at which the bushing may be safely opened. Generally, this will be between 15 °C (59 °F) and 35 °C (95 °F). Pressurized hot oil may suddenly gush from the fill plug if it is removed while at elevated temperature, causing burn hazards. Generally, the bushing will be a little cooler than the top oil temperature, so this temperature gauge may be used as a guide if the gauge has been tested as mentioned in section 4.3.

About 90% of all preventable bushing failures are caused by moisture entering through leaky gaskets, cracks, or seals. Internal moisture can be detected by Doble testing. See *FIST 3-2* [10] and *Doble Bushing Field Test Guide* [9] for troubles and corrective actions. Internal moisture causes deterioration of the insulation of the bushing and can result in explosive failure, causing extensive transformer and other equipment damage, as well as hazards to workers.

After 1 month of service and yearly, examine the bushings with an IR camera [4, 8]; if one phase shows a markedly higher temperature, there is probably a bad connection. The connection at the top is usually the poor one; however, a bad connection inside the transformer tank will usually show a higher temperature at the top as well. In addition, a bad connection inside the transformer will usually show hot metal gases (ethane and ethylene) in the DGA.

Once every 3 to 5 years, perform a close physical inspection and cleaning of the bushings [10]. Check carefully for leaks, cracks, and carbon tracking. This inspection will be required more often in atmospheres where salts and dust deposits appear on the bushings. In conditions that produce deposits, a light application of Dow Corning grease DC-5 or GE Insulgel will help reduce risk of external flashover. The disadvantage of this treatment is that a grease buildup may occur. In high humidity and wet areas, a better choice may be a high-quality silicone paste wax applied to the porcelain, which will reduce the risk of flashover. A spray-on wax containing silicone, such as Turtle Wax brand, has been found to be very useful for cleaning and waxing in one operation, providing the deposits are not too hard. Wax will cause water to form beads, rather than a continuous sheet, which reduces flashover risk. Cleaning may involve just spraying with Turtle Wax and wiping with a soft cloth. A lime removal product, such as "Lime Away," also may be useful. More stubborn contaminants may require solvents, steel wool, and brushes. A high pressure water stream may be required to remove salt and other water soluble deposits. Limestone powder blasting with dry air will safely remove metallic oxides, chemicals, salt-cake, and almost any hard contaminant. Other materials, such as potter's clay, walnut or pecan shells, or crushed coconut shells, are also used for hard contaminants. Carbon dioxide (CO_2) pellet blasting is more expensive but virtually eliminates cleanup because it evaporates. Ground-up corn-cob blasting will remove soft pollutants, such as old coatings of built-up grease. A competent, experienced contractor should be employed, and a thorough, written job hazard analysis (JHA) should be performed when any of these treatments are used.

Corona (air ionization) may be visible at the tops of bushings at twilight or night, especially during periods of rain, mist, fog, or high humidity. At the top, corona is considered normal; however, as a bushing becomes more and more contaminated, corona will creep lower and lower. If the bushing is not cleaned, flashover will occur when corona nears the grounded transformer top. If corona seems to be lower than the top of the bushing, inspect, Doble test, and clean the bushing as quickly as possible. If flashover occurs (phase to ground fault), it could destroy the bushing and cause an extended outage. Line-to-line faults also can occur if all the bushings are contaminated and flashover occurs. A corona scope may be used to view and photograph low levels of corona indoors under normal illumination and outdoors at twilight or night. High levels of corona may be viewed outdoors in the daytime if a dark background is available, such as trees, canyon walls, buildings, etc. The corona scope design is primarily for indoor and night use; it cannot be used with blue or cloudy sky background. This technology is available at the Technical Service Center (TSC), D-8450.

Once every 3 to 5 years, depending on the atmosphere and service conditions, the bushings should be Doble tested. Refer to Doble M-4000 test set instructions, *Doble Bushing Field Test Guide* [9], *FIST 3-2* [10], and the manufacturer's instructions for proper values and test procedures. Bushings should be cleaned prior to Doble testing. Contamination on the insulating surface will cause the results to be inaccurate. Testing may also be done before and after cleaning to check methods of cleaning. As the bushings age and begin to deteriorate, reduce the testing interval to 1 year. Keep accurate records of results so that replacements can be ordered in advance, before you have to remove bushings from service.

CAUTION:

See the transformer manual for detailed instructions on
cleaning and repairing your specific bushing surfaces.
Different solvents, wiping materials, and cleaning methods
may be required for different bushings. Different repair
techniques may also be required for small cracks and chips.
Generally, glyptal or insulating varnish will repair small
scratches, hairline cracks, and chips. Sharp edges of a chip
should be honed smooth, and the defective area should be
painted with insulating varnish to provide a glossy finish.
Hairline cracks in the surface of the porcelain must be sealed
because accumulated dirt and moisture in the crack may
result in flashover. Epoxy should be used to repair larger
chips. If a bushing insulator has a large chip that reduces
the flashover distance or has a large crack totally through the
insulator, the bushing must be replaced. Some
manufacturers offer repair service to damaged bushings that
cannot be repaired in the field. Contact the manufacturer for
your particular bushings if you have repair questions.

4.9 Oil Preservation Sealing Systems

The purpose of sealing systems is to prevent air and moisture from
contaminating oil and cellulose insulation. Sealing systems are
designed to prevent oil inside the transformer from coming into
contact with air. Air contains moisture, which causes sludging and an
abundant supply of oxygen. Oxygen, in combination with moisture,
causes greatly accelerated deterioration of the cellulose. This oxygen-
moisture combination will greatly reduce service life of the
transformer.

Sealing systems on many existing Reclamation power transformers are
of the inert gas (nitrogen) pressure design; however, Reclamation has
many other designs. Current practice is to buy only conservator
designs with bladders for transformer voltages 115 kV and above and
capacities above 10 MVA. Below these values, we buy only inert gas
pressure system transformers, as depicted in figure 36 (page 79).

Some of the sealing systems are explained below. There may be variations of each design, and not every design is discussed. Early sealing system types are discussed first, followed by more modern types.

4.9.1 Sealing Systems Types

4.9.1.1 Free Breathing – Sealing systems have progressed from early designs of "free breathing" tanks, in which an air space on top of the oil is vented to atmosphere through a breather pipe. The pipe typically is screened to keep out insects and rodents and turned down to prevent rain from entering. Breathing is caused by expansion and contraction of the oil as temperature changes. These earlier designs do not use an air dryer, and condensation from moisture forms on inside walls and tank top. Moisture, oxygen, and nitrogen would also dissolve directly into oil from the air. This is not the best design. As mentioned before, a combination of oxygen and moisture accelerates deterioration of cellulose insulation. Moisture also decreases dielectric strength, which destroys insulating quality of the oil and causes sludge to form. If you have one or more of these earlier design transformers, it is recommended that a desiccant type air dryer be added to the breather pipe.

4.9.1.2 Sealed or Pressurized Breathing – This design is similar to the free breathing type with the addition of a pressure/vacuum bleeder valve. When the transformer was installed, pressurized dry air or nitrogen was placed on top of the oil. The bleeder valve is designed to hold pressure inside to approximately plus or minus 5 psi (figure 35). The same problems with moisture and oxygen occur as previously described. However, these problems are not as severe because "breathing" is limited by the bleeder valve. Air or nitrogen (N_2) is exhausted to the outside atmosphere when a positive pressure greater than 5 psi occurs inside the tank. This process does not add moisture and oxygen to the tank. However, when cooling, the oil contracts and, if pressure falls 5 psi below the outside atmosphere, the valve allows outside air into the tank, which pulls in moisture and oxygen.

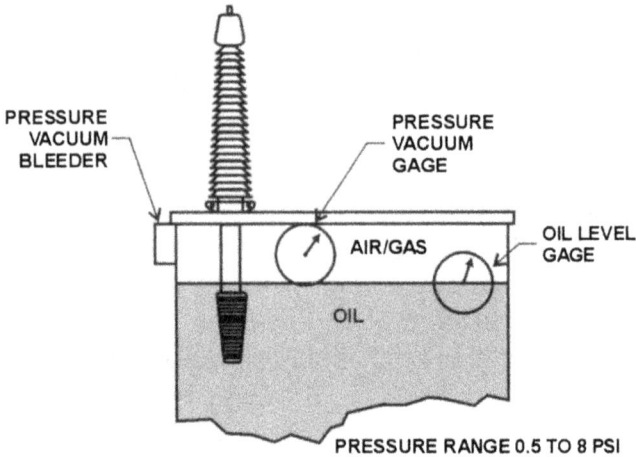

Figure 35 – Pressurized Breathing Transformer.

Once each year, check the pressure gauge against the weekly data sheets; if the pressure never varies with seasonal temperature changes, the gauge is defective. Add nitrogen if the pressure falls below 1 psi to keep moisture laden air from being pulled in. Add enough N_2 to bring the pressure to between 2 and 4 psi.

4.9.1.3 Pressurized Inert Gas Sealed System – This system keeps space above the oil pressurized with a dry inert gas, normally nitrogen (figure 36). This design prevents air and moisture from coming into contact with insulating oil. Pressure is maintained by a nitrogen gas bottle with the pressure regulated normally between 0.5 and 5 psi. Pressure gauges are provided in the nitrogen cubicle for both high and low pressures (figure 37). A pressure/vacuum gauge is normally connected to read low pressure gas inside the tank. This gauge may be located on the transformer and normally has high-pressure and low-pressure alarm contacts. See section 4.9.2, which follows.

If the transformer has a nitrogen blanket, check the pressure gauge for proper pressure. Look at the operators recording of pressures from the pressure gauge. If this does not change, the gauge is probably

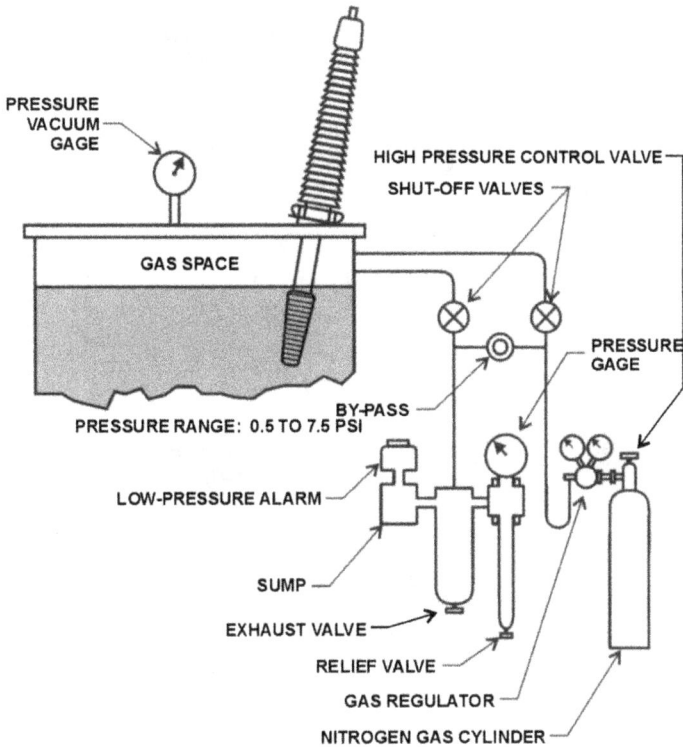

Figure 36 – Pressurized Inert Gas Transformer.

CAUTION:

When replacing nitrogen cylinders, do not just order a "nitrogen cylinder" from the local welding supplier. Nitrogen for transformers should meet American Society for Testing and Materials (ASTM) D-1933 Type III with -59 °C dew point as specified in IEEE C-57.12.00-1993, paragraph 6.6.3 [29, 2].

defective. Check the nitrogen bottle to insure the nitrogen is the proper quality (see *Power Equipment Bulletin No. 5* [40]). Check for any increased usage of nitrogen which indicates a leak. Smaller

PRESSURE-VACUME GAGE
FOR TRANSFORMER GAS
SPACE WITH HIGH- AND LOW-
PRESSURE ALARM CONTACTS

ALARM CONTACT
ADJUSTING SCREW
(LOW-PRESSURE)

TO GAS
SPACE IN
TRANSFORMER

HIGH-PRESSURE
GAGE WITH ALARM
CONTACTS FOR
LOW CYLINDER
PRESSURE

ALARM CONTACT
ADJUSTING SCREW
(HIGH-PRESSURE)

BY-PASS VALVE
(NORMALLY
CLOSED)

HIGH-PRESSURE
REGULATOR WITH
ADJUSTING SCREW

LOW-PRESSURE
REGULATOR WITH
PRESSURE
BLEEDER DEVICE

INLET
CONNECTION
NUT

BY-PASS
LINE

SUMP

SUMP DRAIN
PLUG

Front View   **Side View**

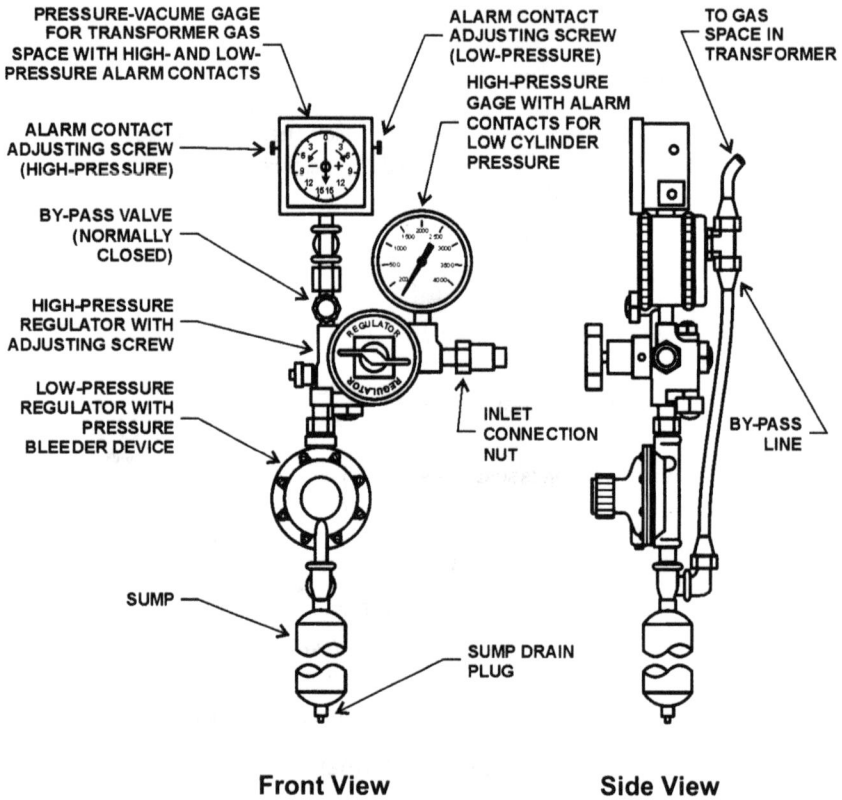

Figure 37 – Gas Pressure Control Components.

transformers such as station service or smaller generator-step-up transformers may not have nitrogen bottles attached to replace lost nitrogen. Be especially watchful of the pressure gauge and the operator's records of pressures with these. The pressure gauge can be defective for years, and no one will notice. The gauge will read nearly the same and will not vary much over winter and summer or night and day. Meanwhile, a nitrogen leak can develop, and all the N_2 will be lost. This allows air with oxygen and moisture to enter and deteriorate the oil and insulation. Watch for increased oxygen and moisture in the DGA. An ultrasonic and sonic leak detection instrument (P-2000) is used for locating N_2 leaks. Soap bubbles also may be used.

4.9.2 Gas Pressure Control Components

After 1 month of service and yearly, inspect the gas pressure control components. There is normally an adjustable, three-element, pressure control system for inert gas, which maintains a pressure range of 0.5 to 5 psi in the transformer tank. There is also a bleeder valve that exhausts gas to atmosphere when pressure exceeds relief pressure of the valve, normally 5 to 8 psi.

CAUTION:

The component part descriptions below are for the typical, three-stage pressure regulating equipment supplying inert gas to the transformer. Your particular unit may be different, so check your transformer instruction manual.

4.9.2.1 High-Pressure Gauge – The high-pressure gauge is

attached between the nitrogen cylinder and high-pressure regulator that indicates cylinder pressure. When the cylinder is full, the gauge will read approximately 2,400 psi. Normally, the gauge will be equipped with a low-pressure alarm that activates when the cylinder is getting low (around 500 psi). However, gas will still be supplied, and the regulating equipment will continue to function until the cylinder is empty. Refer to figure 37 for the following descriptions.

4.9.2.2 High-Pressure Regulator – The high-pressure

regulator has two stages. The input of the first stage is connected to the cylinder, and the output of the first stage is connected internally to the input of the second stage. This holds output pressure of the second stage constant. The first stage output is adjustable by a hand-operated lever and can deliver a maximum pressure equal to the pressure in the cylinder (2,400 psi when full) down to zero. The second stage output is varied by turning the adjusting screw, normally adjusted to supply approximately 10 psi to the input of the low-pressure regulator.

4.9.2.3 Low-Pressure Regulator – The low-pressure regulator is the third stage and controls pressure and flow to the gas space of the transformer. The input of this regulator is connected to the output of the second stage (approximately 10 psi). This regulator is typically set at the factory to supply gas to the transformer at a pressure of approximately 0.5 psi, and it needs no adjustment. If a different pressure is required, the regulator can be adjusted by varying spring tension on the valve diaphragm. Pressure is set at this low value because major pressure changes inside the transformer result from expansion and contraction of oil. The purpose of this gas feed is to make up for small leaks in the tank gaskets and elsewhere so that air cannot enter. Typically, a spring-loaded bleeder for high-pressure relief is built into the regulator and is set at the factory to relieve pressures in excess of 8 psi. The valve will close when pressure drops below the setting, preventing further loss of gas.

4.9.2.4 Bypass Valve Assembly – The bypass valve assembly opens a bypass line around the low-pressure regulator and allows the second stage of the high-pressure regulator to furnish gas directly to the transformer. The purpose of this assembly is to allow much faster filling/purging of the gas space during initial installation or when the transformer tank needs to be refilled after being opened for inspection.

CAUTION:

During normal operation, the bypass valve must be closed,
or pressure in the tank will be too high.

4.9.2.5 Oil Sump – The oil sump is located at the bottom of the pressure regulating system between the low-pressure regulator and shutoff valve C. The sump collects oil and/or moisture that may have condensed in the low-pressure fill line. The drain plug at the bottom of the sump should be removed before the system is put into operation, as well as once each year during operation, to drain any residual oil in the line. This sump and line will be at the same pressure as the gas space in the top of the transformer. The sump should always be at a safe pressure (less than 10 psi) so the plug can be removed to allow the line to purge a few seconds and blow out the oil. However, always look at the gas space pressure gauge on the transformer or the low-pressure gauge in the nitrogen cabinet, just to be sure, before removing the drain plug.

4.9.2.6 Shutoff Valves – The shutoff valves are located near the top of the cabinet for the purpose of isolating the transformer tank for shipping or maintenance. These valves are normally of double-seat construction and should be fully opened against the stop to prevent gas leakage around the stem. A shutoff valve is also provided for the purpose of shutting off the nitrogen flow to the transformer tank. This shutoff valve must be closed prior to changing cylinders to keep the gas in the transformer tank from bleeding off.

4.9.2.7 Sampling and Purge Valve – The sampling and purge valve is normally located in the upper right of the nitrogen cabinet. This valve is typically equipped with a hose fitting; the other side is connected directly to the transformer gas space by copper tubing. This valve is opened while purging the gas space during a new installation or maintenance refill, and it provides a path to exhaust air as the gas space is filled with nitrogen. This valve is also opened when a gas sample is taken from the gas space for analysis. When taking gas samples, the line must be sufficiently purged so that the sample will be from gas above the transformer oil and not just gas in the line. This valve must be tightly closed during normal operation to prevent gas leakage.

4.9.2.8 Free Breathing Conservator – This design adds an expansion tank (conservator) above the transformer so that the main tank may be completely filled with oil (figure 38). Oil expansion and air exchange with the atmosphere (breathing) occur away from the oil in the transformer. This design reduces oxygen and moisture

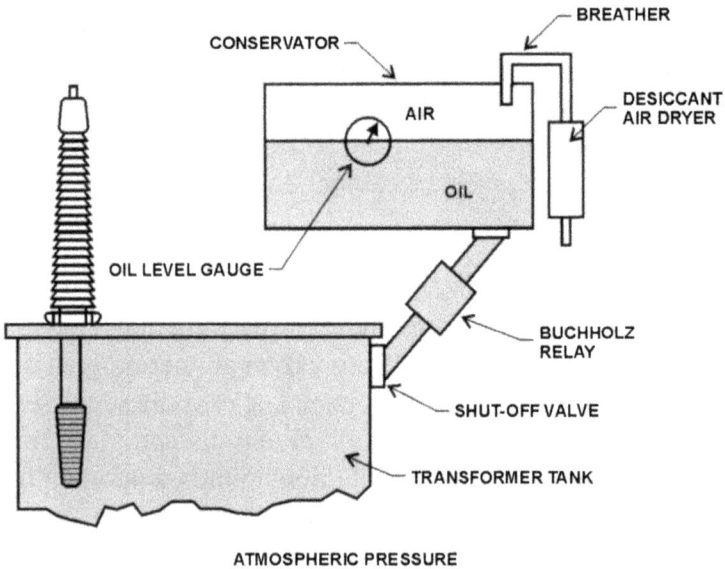

Figure 38 – Free Breathing Conservator.

contamination because only a small portion of oil is exchanged between the main tank and conservator. An oil/air interface still exists in the conservator, exposing the oil to air. Eventually, oil in the conservator is exchanged with oil in the main tank, and oxygen and other contaminants gain access to the insulation.

If you have transformers of this design, it is recommended that a bladder or diaphragm-type conservator be installed (described below) or retrofitted to the original conservator. In addition, a desiccant-type air dryer should also be installed.

4.9.2.9 Conservator with Bladder or Diaphragm Design – A conservator with bladder or diaphragm (figure 39) is similar to the design in figure 38 with an added air bladder (balloon) or flat diaphragm in the conservator. The bladder or diaphragm expands and contracts with the oil and isolates it from the atmosphere. The inside of the bladder or top of the diaphragm is open to atmospheric pressure through a desiccant air dryer. As oil expands and contracts and as

84

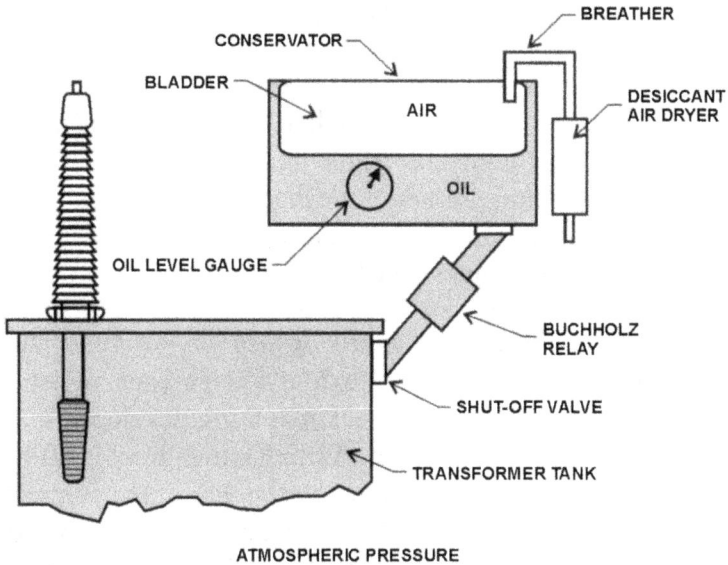

Figure 39 – Conservator with Bladder.

NOTE:

A vacuum will appear in the transformer if piping between the air dryer and conservator is too small; if the air intake to the dryer is too small, or if the piping is partially blocked. The bladder cannot take in air fast enough when the oil level is decreasing due to rapidly falling temperature. Minimum ¾- to 1-inch piping is recommended. This problem is especially prevalent with transformers that are frequently in and out of service and located in geographic areas with large temperature variations. This situation may allow bubbles to form in the oil and may even activate gas detector relays, such as the Buchholz and/or bladder failure relay. The vacuum may also pull in air around gaskets that are not tight enough or that have deteriorated (which may also cause bubbles) [4].

---

atmospheric pressure changes, the bladder or diaphragm "breathes" air in and out. This keeps air and transformer oil essentially at atmospheric pressure. The oil level gauge on the conservator typically is magnetic, like those mentioned earlier, except the float is positioned near the center of the underside of the bladder. With a diaphragm, the level indicator arm rides on top of the diaphragm. Examine the air dryer periodically and change the desiccant when approximately one-third of the material changes color.

4.9.2.9.1 Conservator Inspection – If atmospheric gases (nitrogen, oxygen, carbon dioxide) and, perhaps, moisture increase suddenly in the DGA, a leak may have developed in the conservator diaphragm or bladder. With the transformer offline and under clearance, open the inspection port on top of the conservator and look inside with a flashlight. If there is a leak, oil will be visible on top of the diaphragm or inside the bladder. Re-close the conservator and replace the bladder or diaphragm at the first opportunity by scheduling an outage. If there is no gas inside the Buchholz relay, the transformer may be re-energized after bleeding the air out of the bladder failure relay. A DGA should be taken immediately to check for oxygen (O_2), N_2, and moisture. However, the transformer may be operated until a new bladder is installed, keeping a close eye on the DGAs. It is recommended that DGAs be performed every 3 months until the new bladder is installed. After the bladder installation, the oil may need to be de-gassed if O_2 exceeds 10,000 parts per million (ppm). Also, carefully check the moisture level in the DGAs to ensure it is below recommended levels for the particular transformer voltage. Check the desiccant in the breather often; never let more than two-thirds of the desiccant become discolored before renewing it. All efforts must be made to keep the oxygen level below 2,000 ppm and moisture as low as possible.

4.9.2.9.2 Conservator Breather Inspection – Check the dehydrating (desiccant) breather for proper oil level if it is an oil-type unit. Check the color of the desiccant and replace it when approximately one-third remains with the proper color. See figure 40 for a modern oil-type desiccant breather. Notice the pink desiccant at the bottom of the blue indicating that this portion is water saturated.

Notice also that oil is visible in the very bottom 1 inch or so of the unit. Many times, the oil is clear, and the oil level will not be readily apparent. Normally, there is a thin line around the breather near the bottom of the glass; this indicates where the oil level should be. Compare the oil level with the level indicator line and refill, if necessary. Note the 1¼-inch pipe going from the breather to the conservator. Small tubing (½ inch or so) is not large enough to admit

Figure 40 – Conservator Breather.

air quickly when the transformer is de-energized in winter. A transformer can cool so quickly that a vacuum can be created from oil shrinkage with enough force to puncture a bladder. When this happens, the bladder is destroyed; and air is pulled into the conservator creating a large bubble.

4.9.2.9.3 Bladder Failure (Gas Accumulator) Relay. The bladder failure relay (not on diaphragm-type conservators) (figure 41) is mounted on top of the conservator for the purpose of

Figure 41 – Photograph of a Bladder Failure Relay.

detecting air bubbles in the oil. The relay will also serve as a backup to the Buchholz relay. If the Buchholz relay overfills with gas and fails to activate an alarm or shut down, gas will bypass the Buchholz and migrate up into the conservator, eventually to the bladder failure relay. Of course, these gases should also show up in the DGA. However, DGAs are normally taken only once per year, and a problem may not be discovered before these alarms are activated. Figure 42 shows a modern relay. Check your transformer instruction manual for specifics because designs

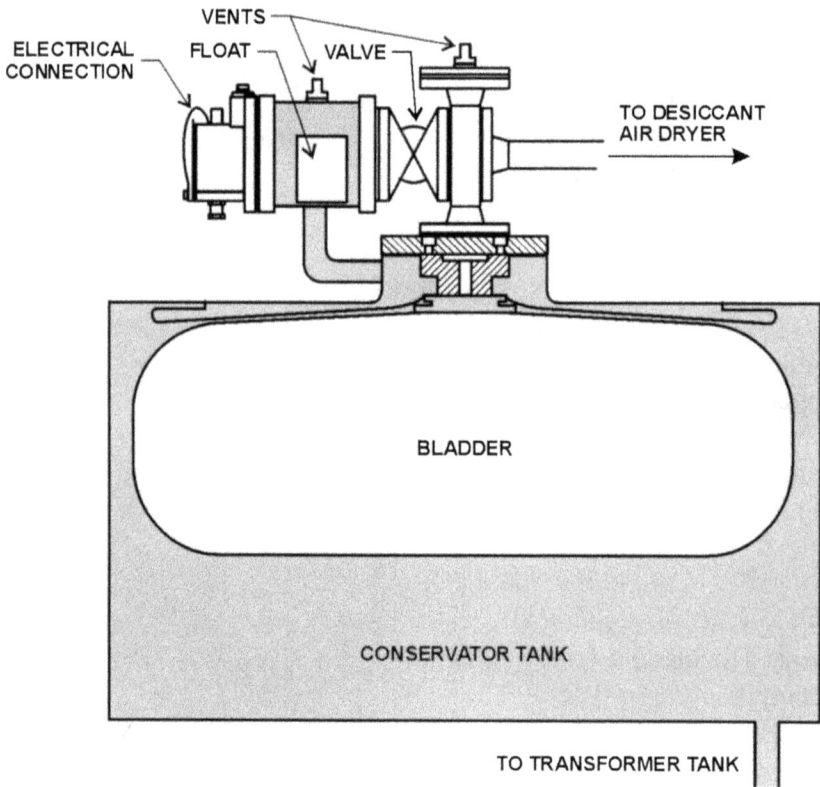

Figure 42 – Bladder Failure Relay.

vary with manufacturers. No bladder is totally impermeable, and a little air will migrate into the oil.

If a hole forms in the bladder, allowing air to migrate into the oil, the relay will detect it. As air rises and enters the relay, oil is displaced and the float drops, activating the alarm. It is similar to the top chamber of a Buchholz relay, since it is filled with oil and contains a float switch.

Every 3 to 5 years usually during Doble testing or if the bladder failure alarm is activated (if the conservator has a diaphragm), place the transformer under clearance and check the Buchholz relay for gas, as mentioned in section 4.7. Open the conservator inspection port and look inside with a flashlight.

Bleed the air/gas from the conservator using the bleed valve on top of the conservator. If the transformer is new and has been in service for only a few months, the problem most likely is air escaping from the structure. With the transformer under clearance, open the inspection port on top of the conservator and look inside the bladder with a flashlight. If oil is found inside the bladder, it has developed a leak; a new one must be ordered and installed.

CAUTION:

Never open the vent of the bladder failure relay unless you have vacuum or pressure equipment available. The oil will fall inside the relay and conservator and pull in air from the outside. You will have to recommission the relay by valving off the conservator and pressurizing the bladder or by placing a vacuum on the relay. See your specific transformer instruction manual for details.

CAUTION:

When the transformer, relay, and bladder are new, some air or gas is normally entrapped in the transformer and piping and takes some time to rise and activate the relay. Do not assume the bladder has failed if the alarm activates within 2 to 3 months after it is put into operation. If this occurs, you will have to recommission the relay with pressure or vacuum. See your specific transformer instruction manual for details. If no more alarms occur, the bladder is intact. If alarms continue, look carefully for oil leaks in the conservator and transformer. An oil leak is usually also an air leak. This may be checked by looking at the nitrogen and oxygen in the dissolved gas analysis. If these gases are increasing, there is probably a leak; with a sealed conservator, there should be little of these gases in the oil. Nitrogen may be high if the transformer was shipped new and was filled with nitrogen.

4.10 Auxiliary Tank Sealing System

The auxiliary tank sealing system incorporates an extra tank between the main transformer tank and the conservator tank. Inert gas (normally nitrogen) is placed above oil in both the main and middle tanks. Only oil in the top conservator tank is exposed to air. A desiccant air dryer may, or may not, be included on the breather. As oil in the main tank expands and contracts with temperature, gas pressure varies above the oil in both (figure 43).

Changes in gas pressure cause oil to go back and forth between the middle tank and the conservator. Air containing oxygen and moisture is not in contact with oil in the main transformer tank. Oxygen and moisture are absorbed by oil in the conservator tank and interchanged with oil in the middle tank. However, since gas in the middle tank interchanges with gas in the main tank, small amounts of oxygen and moisture carried by gas still make their way into the transformer.

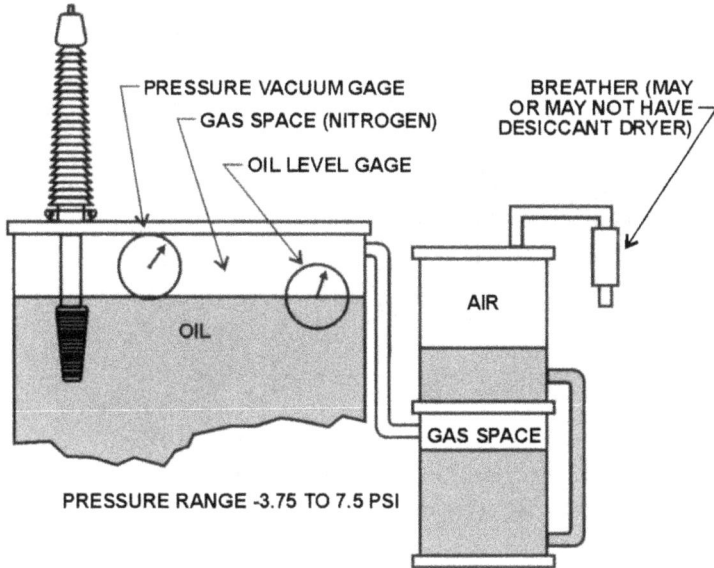

Figure 43 – Auxiliary Sealing System.

With this arrangement, the conservator does not have to be located above the main tank, which reduces the overall height. If you have one or more of these type transformers without desiccant air dryers, they should be installed.

5. Gaskets

Gaskets have several important jobs in sealing systems [7]. A gasket must create a seal and hold it over a long period of time. It must be impervious and not contaminate the insulating fluid or gas above the fluid. It should be easily removed and replaced. A gasket must be elastic enough to flow into imperfections on the sealing surfaces. It must withstand high and low temperatures and remain resilient enough to hold the seal, even with joint movement from expansion, contraction, and vibration. It must be resilient enough to avoid taking a "set," even when exposed for a long time to pressure applied with bolt torque and temperature changes. The gasket must have sufficient strength to resist crushing under applied load and resist blowout under system pressure or vacuum. It must maintain its integrity while being handled or installed. If a gasket fails to meet any of these criteria, a leak will result. Gasket leaks result from improper torque, choosing the wrong type of gasket material, or choosing the wrong size gasket. Improper sealing surface preparation or the gasket taking a "set" (becoming hard and losing its resilience and elasticity) will also cause a leak. Usually, gaskets take a set as a result of temperature extremes and age.

5.1 Sealing (Mating) Surface Preparation

Clean the metal surface thoroughly. Remove all moisture, oil and grease, rust, etc. A wire brush and/or solvent may be required.

CAUTION:

Take extra care that rust and dirt particles never fall into the transformer. The results could be catastrophic when the transformer is energized.

After rust and scale have been removed, metal surfaces should be coated with Loctite Master gasket No. 518. This material will cure after you bolt up the gasket, so additional glue is not necessary. If the temperature is 50 °F or more, you can bolt up the gasket immediately. This material comes in a kit (part No. 22424) with primer, a tube of material, and instructions. If these instructions are followed, the seal will last many years, and the gasket will be easy to remove later, if necessary. If the temperature is under 50 °F, wait about ½ to 1 hour after applying the material to surfaces before bolting. If you are using cork-nitrile or cork-neoprene, you can also seal gasket surfaces (including the edge of the gasket) with this same material. Loctite makes other sealers that can be used to seal gaskets, such as "Hi-tack."

GE glyptol No. 1201B-red can also be used to paint gasket and metal surfaces; but it takes more time, and you must be more cautious about temperature. If possible, this work should be done in temperatures above 70 °F to speed paint curing. Allow the paint to completely dry before applying glue or the new gasket. It is not necessary to remove old glyptol, or other primer, or old glue if the surface is fairly smooth and uniform.

Choose the correct replacement gasket. The main influences on gasket material selection are design of the gasket joint, maximum and minimum operating temperature, type of fluid contained, and internal pressure of the transformer.

CAUTION:

Most synthetic rubber compounds, including nitrile (Buna N), contain some carbon, which makes it semiconductive. Take extra care to never drop a gasket or pieces of gasket into a transformer tank. The results could be catastrophic when the transformer is energized.

5.2 Cork-Nitrile

Cork-nitrile should be used if the joint **does not have grooves** or limits. This material performs better than cork-neoprene because it does not take a set as easily and conforms better to mating surfaces. It also performs better at higher temperatures. Be extra careful when you store this material because it looks like cork-neoprene (described in section 5.3), and they easily are mistaken for each other. Compression is the same as for cork-neoprene, about 45%. Cork-nitrile should recover 80% of its thickness with compression of 400 psi in accordance with ASTM F36. Hardness should be 60 to 75 durometer in accordance with ASTM D2240. (See published specifications for E-98 by manufacturer Dodge-Regupol Inc., Lancaster, Pennsylvania.)

CAUTION:

Cork nitrile has a shelf life of only about 2 years, so do not
order and stock more than can be used during this time.

5.3 Cork-Neoprene

Cork-neoprene mixture (called coroprene) can also be used; however,
it does not perform as well as cork-nitrile. This material takes a set
when it is compressed and should only be used when there are no
expansion limiting grooves. Using cork-neoprene in grooves can
result in leaks from expansion and contraction of mating surfaces. The
material is very porous and should be sealed on both sides and edges
with a thin coat of Glyptol No. 1201B red or similar sealer before
installation. Glyptol No. 1201B is a slow drying paint used to seal
metal flanges and gaskets, and the paint should be allowed to dry
totally before installation. Once compressed, this gasket should never
be reused. These gaskets should be kept above 35 °F before
installation to prevent them from becoming hard. Gaskets should be
cut and sealed (painted) indoors at temperatures above 70 °F for ease
of handling and to reduce paint curing time. Avoid installing cork-
neoprene gaskets when temperatures are at or near freezing because
the gasket could be damaged and leak. Cork-neoprene gaskets must be
evenly compressed at about 43 to 45%. For example, if the gasket is
¼-inch thick, 0.43 x 0.25 = 0.10. When the gasket is torqued down, it
should be compressed about 0.10 inch. Or you may subtract 0.1 from
¼ inch to calculate the thickness of the gasket after it is compressed.
In this case, ¼ = 0.25 so 0.25 minus 0.10 = 0.15 inch would be the
final distance between the mating surfaces after the gasket is
compressed. In an emergency, if compression limits are required on
this gasket, split lock washers may be used. Bend the washers until
they are flat and install enough of them (minimum of three), evenly
spaced, in the center of the gasket cross section to prevent excessive

compression. The thickness of the washers should be such that the gasket compression is limited to approximately 43%, as explained above.

5.4 Nitrile "NBR"

Nitrile "NBR" (buna N) with 50 to 60 duro (hardness) is generally the material that should be chosen for most transformer applications.

CAUTION:

Do not confuse this material with butyl rubber. Butyl is not a satisfactory material for transformer gaskets. The terms butyl and buna are easily confused, and care must be taken to make sure nitrile (buna N) is always used and never butyl.

Replace all cork neoprene gaskets with nitrile **if the joint has recesses or expansion limiting grooves**. Be careful to protect nitrile from sunlight; it is not sunlight resistant and will deteriorate, even if only the edges are exposed. It should not be greased when it is used in a nonmovable (static) seal. When joints have to slide during installation or are used as a moveable seal (such as bushing caps, oil cooler isolation valves, and tap changer drive shafts), the gasket or O-ring should be lubricated with a thin coating of DOW No. 111, No. 714, or equivalent grease. These are very thin and provide a good seal. Nitrile performs better than cork-neoprene; when exposed to higher temperatures, it will perform well up to 65 °C (150 °F).

5.4.1 Viton

Viton should be used only for gaskets and O-rings in temperatures higher than 65 °C or for applications requiring motion (shaft seals, etc.). Viton is very tough and wear resistant; however, it is very expensive ($1,000+ per sheet). Therefore, it should not be used unless it is needed for high wear or high temperature applications. Viton should only be used with compression limiter grooves and recesses.

Store nitrile and viton separately, or order them in different colors; the materials look alike and can be easily confused, causing a much more expensive gasket to be installed unnecessarily. Compression and fill requirements for Viton are the same as those for nitrile, as outlined above and shown in table 3.

Table 3 – Transformer Gasket Application Summary

Gasket Material	Best Temperature Range	Percent Compar- ison	Compatible Fluids	UV Resist	Best Applications
Neoprene Use nitrile, except where there is ultraviolet (UV) exposure, or use viton.	-54 to 60 °C (-65 to 140 °F) not good with temperature swings	30 to 33	Askarels and hydrocarbon fluids	Yes	Use only with compression limits or recesses and use only if UV resistance is needed.
Cork-Neoprene (Coroprene) This material takes a set easily.	0 to 60 °C (32 to 140 °F)	40	Mineral oil R-Temp Alpha 1	No	Use only for flat to flat surface gaskets with no grooves or compression limits.
Cork-Nitrile (best) does not take a set as easily as cork-neoprene.	-5 to 60 °C (23 to 140 °F)	40	Mineral oil R-Temp Alpha 1	No	Use only for flat to flat surface gaskets with no grooves or compression limits.
Nitrile (Buna N) Use this except in high temperature, high wear, or UV.	-5 to 65 °C (23 to 150 °F)	25 to 50	Mineral oil R-Temp, Alpha 1 Excellent for Hydrocarbon fluids	No	O-rings, flat and extruded gaskets; use with compression limiters or recess only.[1]
Viton Use for high wear and high temperature applications.	-20 to 150 °C (-4 to 302 °F)	30 to 33	Silicone, Alpha 1 Mineral oil	Yes	High temperature; O-rings, flat and extruded gaskets; use with compression limiter groove or recess.[2]

[1] Nitrile (buna N) can also be used in low wear applications and temperatures less than 65 °C.

[2] Viton O-rings are best for wear resistance and tolerating temperature variations.

5.5 Gasket Sizing for Standard Groove Depths

Nitrile is the chosen example because it is the most commonly used material for transformer gasketing. As shown in table 3, nitrile compression should be 25 to 50%. Nitrile sheets are available in 1/16-inch-thick increments.

Gasket thickness is determined by groove depth and standard gasket thickness. Choose the sheet thickness so that ¼ to ⅓ thickness of the gasket will protrude above the groove; this is the amount available to be compressed (see table 4). Gasket sheets come in standard thicknesses in 1/16-inch increments. Choose one that allows a of the gasket to stick out above the groove, if possible, but never choose a thickness that allows less than ¼ or as much as ½ of the gasket to protrude above the groove. Do not try to remove old primer from the groove.

Table 4 – Vertical Groove Compression for Circular Nitrile Gasket

Standard groove depth (in inches)	Recommended gasket thickness (in inches)	Available to compress (in inches)	Available compression (percent)
3/32	1/8	1/32	25
1/8	3/16	1/16	33
3/16	1/4	1/16	25
1/4	3/8	1/8	33
3/8	1/2	1/8	25

Horizontal groove fill is determined by how wide the groove is. The groove width is equal to the outer diameter (OD) minus the inner diameter (ID) divided by two:

$$\frac{OD - ID}{2}$$

Or just measure the groove width with an accurate caliper.

Room left for the gasket to expand while being compressed is equal to the width of the groove minus the width of the gasket. For nitrile, the amount of horizontal room needed is about 15 to 25%. Therefore, you need to cut the gasket cross section so that it fills about 75 to 85% of the width of the groove.

For example, an 8-inch OD groove with a 6-inch ID, $\frac{OD-ID}{2}$, is $\frac{8-6}{2}$ =

1 inch. Therefore, the width of the groove is 1 inch. Because we have to leave 25% expansion space, the width of the gasket is 75% of 1 inch, or ¾ inch. So that the gasket can expand equally toward the center and toward the outside, you should leave one-half the expansion space at the inner diameter of the groove and one-half at the outer. In this example, there should be a total space of 25% of 1 inch or (¼ inch) for expansion after the gasket is inserted, so you should leave ⅛-inch space at the OD and ⅛-inch space at the ID. See figure 44.

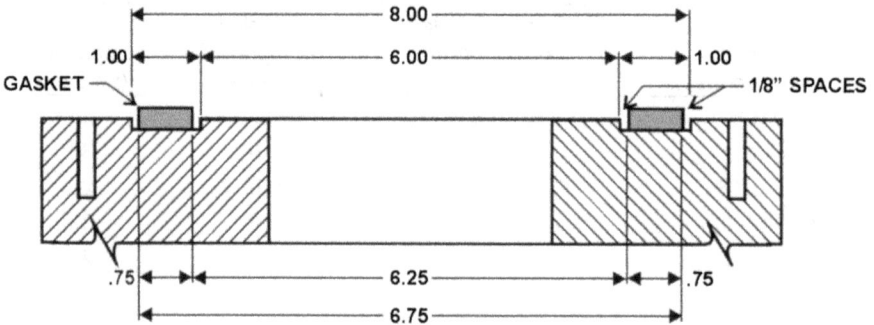

Figure 44 – Cross Section of Circular Gasket in Groove.

Always cut the outer diameter first. In this example, the outer diameter would be 8 inches **minus** ¼ inch, or 7¾ inches.

NOTE:

Since ⅛-inch space is required all around the gasket,
¼ inch must be subtracted to allow ⅛-inch on both sides.
The inner diameter would be 6 inches plus ¼ inch, or
6¼ inches. Note that ¼ inch is subtracted from the OD but
added to the ID.

To check yourself, subtract the inner radius from the outer radius to make sure you get the same gasket width calculated above. In this example, 3⅞ inches (outer radius, ½ of 7¾), minus 3⅛ inches (inner radius, ½ of 6¼), is ¾ inch, which is the correct gasket width.

5.6 Rectangular Nitrile Gaskets

Rectangular nitrile gaskets larger than sheet stock on hand can be fabricated by cutting strips and corners with a table saw or a utility knife with razor blade. Applying a little transformer oil or WD-40 oil makes cutting easier. Nitrile is also available in spools in standard ribbon sizes. The ends may be joined using a cyanoacrylate adhesive (super glue). Although there are many types of this glue, only a few of them work well with nitrile. Since all these glues have a very limited shelf life, remember to always keep them refrigerated to extend their shelf life. Lawson Rubber Bonder No. 92081 best withstands temperature changes and compression. The Lawson part number is 90286, and it is available from Lawson Products Co. in Reno, Nevada, (702-856-1381). Loctite 404 also works but does not survive temperature variations as well. It is available at NAPA Auto Parts Stores. Shelf life is critical. Always secure a new supply when a gasketing job is started; **never** use an old bottle that has been on the shelf since the last job.

NOTE:

Maximum horizontal fill of the groove should be 75 to 85%, as explained above in the circular gasket section. However, it is not necessary to fill the groove fully to 75% to obtain a good seal. Choose the width of ribbon that comes close to, but does not exceed, 75 to 80%. If one standard ribbon width fills only 70% of the groove and the next size standard width fills 90%, choose the size that fills 70%. As in the circular groove explained above, place the gasket so that expansion space is equal on both sides. The key point is that the cross-sectional area of the gasket remains the same while the cover is tightened; the thickness decreases, but the width increases. See below and figure 45.

CAUTION:

Nitrile (buna N) is a synthetic rubber compound and, as cover bolts are tightened, the gasket is compressed. Thickness of the gasket is decreased, and the width is increased. If a gasket is too large, rubber will be pressed into the void between the cover and the sealing surface. This will prevent a metal-to-metal seal, and a leak will result. It is best if the cross-sectional area of the gasket is a little smaller than the groove cross-sectional area. As cover bolts are tightened, the thickness of the gasket decreases, but the width increases so that cross-sectional area (thickness times the width) remains the same. Care must be taken to ensure that the gasket cross-sectional area is equal to or slightly smaller (never larger) than the groove cross-sectional area. This will provide space for the rubber to expand in the groove so that it will not be forced out into the metal-to-metal contact area (see figure 45). If it is forced out into the "metal-to-metal" seal area, a leak generally will be the result. When this happens, our first response is to tighten the bolts, which bends the cover around the gasket material in the metal-to-metal contact area. The leak may stop (more often, it will not); but the next time the cover is removed, getting a proper seal is almost impossible because the cover is bent. Take extra care to correctly size the gasket to prevent these problems.

CROSS SECTION OF GASKET REMAINS CONSTANT
BEFORE TIGHTENING AND AFTER.
w X d = gw X GT

TOP COVER

BEFORE TIGHTENING

GASKET
gw
gt

d

w

GROOVE CAVITY

AFTER TIGHTENING - BEST SEAL
(GASKET FILLS CAVITY EXACTLY)
DIFFICULT TO ACHIEVE

GASKET

RECOMMENDED FIT - AFTER TIGHTENING - GOOD SEAL
(GASKET SLIGHTLY SMALLER THAN CAVITY EXACTLY)
MUCH EASIER TO ACHIEVE

GASKET

AFTER TIGHTENING - POOR SEAL
(GASKET TOO LARGE - PREVENTING METAL TO METAL CONTACT)
gw X gt LARGER THAN w X d

Gasket

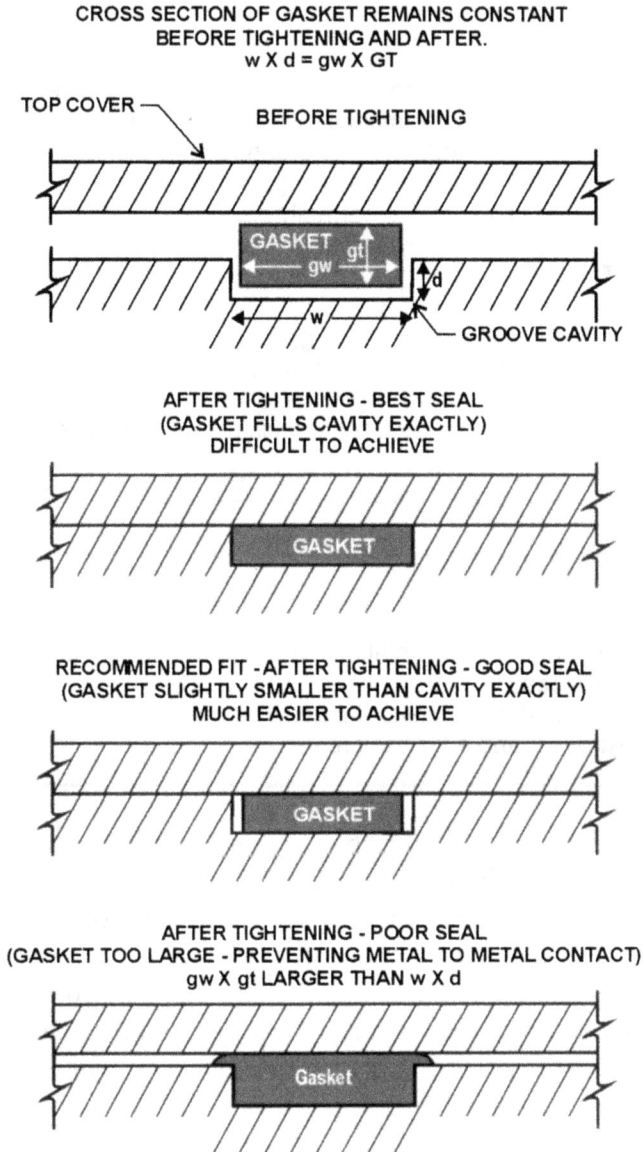

Figure 45 – Cross Section of Gasket Remains Constant
Before Tightening and After. wxd = gw x gt

CAUTION:

On some older bushings used on 15-kV voltages and above, it is necessary to install a semiconductive gasket. This type bushing (such as GE type L) has no ground connection between the bottom porcelain skirt flange and the ground ring. The bottom of the skirt is normally painted with a conductive paint, and then a semiconductive gasket is installed. This allows static electric charges to bleed off to the ground. The gaskets are typically a semiconductive neoprene material. Sometimes, the gasket will have conductive metal staples near the center to bleed off these charges. When replacing this type gasket, always replace with like material. If like gasket material is not available, use cork-neoprene.

NOTE:

Failure to provide a path for static electric charges to get to the ground will result in corona discharges between the ground sleeve and the bushing flange. The gasket will be rapidly destroyed, and a leak will result.

When bonding the ends of ribbon together, the ends should be cut at an angle (scarfed) at about 15 degrees. The best bond occurs when the length of the angle cut is about four times the thickness of the gasket (see table 5). With practice, a craftsperson can cut 15-degree scarfs with a utility knife. A jig can also be made from wood to hold the gasket at a 15-degree angle for cutting and sanding. The ends may be further fine-sanded or ground on a fine bench grinder wheel to match perfectly before applying glue.

Thin metal conductive shim stock may be folded over the outer perimeter around approximately one-half the circumference. These pieces of shim stock should be evenly spaced around the circumference and stick far enough in toward the center so that they will be held when the bolts are tightened. As an example, if the gasket

Table 5 – Vertical Groove Compression for Rectangular Nitrile Gaskets

Standard Groove Depth (inches)	Standard Ribbon Width (inches)	Recommended Gasket Thickness (inches)	Available to Compress (inches)	Available Compression (inches)
3/32	1/4	1/8	1/32	25
1/8	5/16	3/16	1/16	33
3/16	3/8	1/4	1/16	25
1/4	3/4	3/8	1/8	33
3/8	3/4	1/2	1/8	25

is 8 inches in diameter, the circumference would be πD or 3.1416 times 8 inches = 25.13 inches in circumference. Fifty percent of 25.13 is about 12½ inches. Cut 12 strips 1-inch wide and long enough to be clamped by the flange top and bottom when tightened. Fold them over the outside edge of the gasket, leaving a little more than a 1-inch space between, so that the shim stock pieces will be as evenly spaced as possible around the circumference.

5.7 Bolting Sequences to Avoid Sealing Problems

If proper bolt tightening sequences are not followed or if improper torque is applied to the bolts, sealing problems will result (see figure 46). A slight bow in the flange or lid top (exaggerated for illustration) occurs, which applies uneven pressure to the gasket. This bow compromises the seal, and the gasket will eventually leak.

Figure 47 illustrates proper bolting sequences for various type flanges/ covers. Bolt numbers show the correct tightening sequences.

The numbers do not have to be followed exactly; however, the diagonal tightening patterns should be followed. By using proper torque and the illustrated sequence patterns, sealing problems from improper tightening and uneven pressure on the gasket can be avoided. Use a torque wrench and torque bolts according to the head stamp on the bolt. Check the manufacturer's instruction book for pancake gasket torque values.

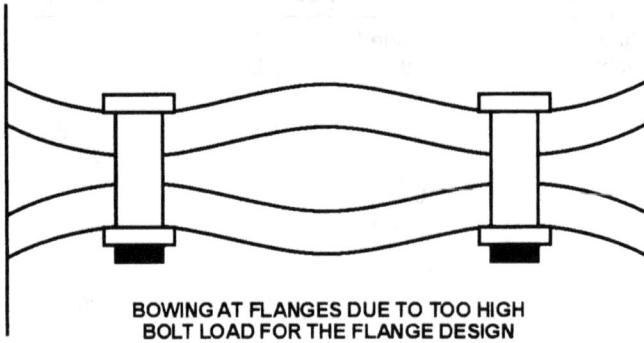

BOWING AT FLANGES DUE TO TOO HIGH
BOLT LOAD FOR THE FLANGE DESIGN

Figure 46 – Bowing at Flanges.

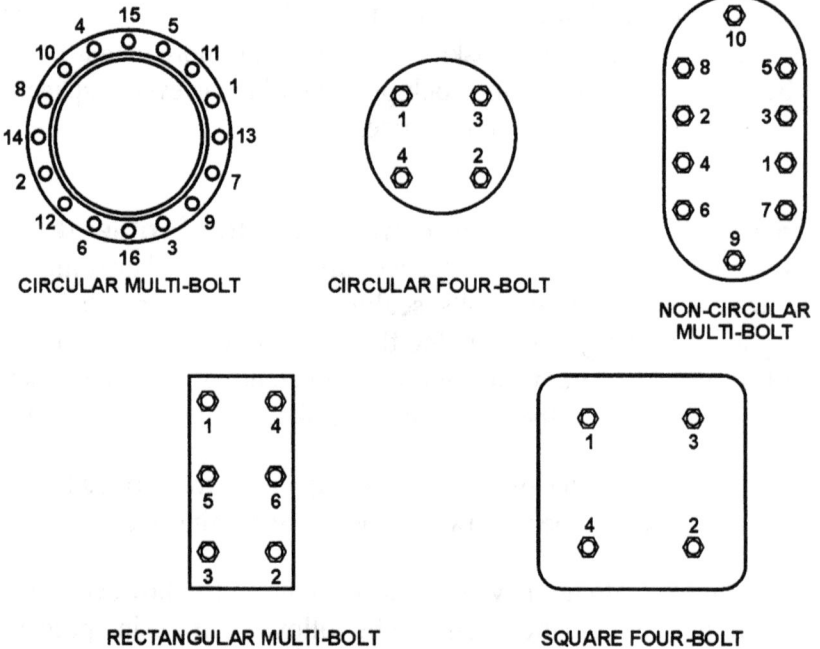

CIRCULAR MULTI-BOLT

CIRCULAR FOUR-BOLT

NON-CIRCULAR
MULTI-BOLT

RECTANGULAR MULTI-BOLT

SQUARE FOUR-BOLT

Figure 47 – Bolt Tightening Sequences.

6. Transformer Oils

6.1 Transformer Oil Functions

Transformer oils perform at least four functions for the transformer. Oil provides insulation, provides cooling, and helps extinguish arcs. Oil also dissolves gases generated by oil degradation, moisture and gas from cellulose insulation, deterioration, and gases and moisture from exposure to atmospheric conditions. Close observation of dissolved gases in the oil, and other oil properties, provides the most valuable information about transformer health. Looking for trends by comparing information provided in several DGAs and understanding its meaning is the most important transformer diagnostic tool.

6.1.1 Dissolved Gas Analysis

After 1 month of service and once each year, and more often if a problem is encountered, complete a DGA. This is, by far, the most important tool for determining the health of a transformer. A DGA is the first indicator of a problem and can identify deteriorating insulation and oil, overheating, hot spots, partial discharge, and arcing. The "health" of the oil is reflective of the health of the transformer itself. A dissolved gas analysis consists of sending transformer oil samples to a commercial laboratory for testing. The most important indicators are the individual and total combustible gas (TCG) generation rates based on the International Electrotechnical Commission (IEC) 60599 [13] and IEEE C 57-104™ [12] standards.

CAUTION:

DGA is unreliable if the transformer is de-energized and has cooled, if the transformer is new, or if it has had less than 1 to 2 weeks of continuous service after oil processing.

This section provides guidance in interpreting DGA and suggests actions based on the analysis. There are no "quick and sure" easy answers when dealing with transformers. Transformers are very complex, very expensive, and very important to Reclamation; and each one is different. Decisions must be based on experienced judgment founded on all available data and consultation with experienced people. Periodic DGA and proper interpretations, along with thorough periodic inspections (covered earlier), are the most important keys to transformer life. Each DGA must be compared to prior DGAs to recognize trends and establish rates of gas generation .

Although examples will be presented later, there are no universally accepted means for interpreting DGA data [16]. Transformers are very complex. Aging, chemical actions and reactions, electric fields, magnetic fields, thermal contraction and expansion, load variations, gravity, and other forces all interact inside the tank. Externally, through-faults, voltage surges, wide ambient temperature changes, and other forces (i.e., the earth's magnetic field and gravity) affect the transformer. There are few, if any, "cut and dried" DGA interpretations; even experts disagree. Consultation with others, experience, study, comparing earlier DGAs, keeping accurate records of a transformer's history, and noting information found when a transformer is disassembled will increase expertise and provide life extension to this critical equipment.

Keeping accurate records of each individual transformer is paramount. If a prior through-fault, overload, cooling problem, or nearby lightning strike has occurred, this information is extremely valuable to determine what is going on inside the transformer. Baseline transformer test information should be established when the transformer is new, or as soon as possible thereafter. This must include DGA, Doble, and other test results (discussed in section 9, "Transformer Testing").

Table 6 represents a summary of DGA transformer analysis. Table 7 is an actual example of a Reclamation transformer.

Table 6 – Transformer DGA Condition Summary Table

	Condition Ranges			
	Good	Fair	Poor	Action
Hydrogen (H₂) ppm	<101	>100 <1,000	>1,000 <2,000	>2,000
Hydrogen (H₂) ppm generation/month	<10	>9 <30	>29 <50	>49
Methane (CH₄) ppm	<121	>120 <401	>400 <1,001	>1,000
Methane (CH₄) ppm generation/month	<8	>7 <23	>22 <38	>37
Ethane (C₂H₆) ppm	<66	>64 <101	>100 <151	>150
Ethane (C₂H₆) ppm generation/month	<8	>7 <23	>22 <38	>37
Ethylene (C₂H₄) ppm	<51	>50 <101	>100 <201	>200
Ethylene (C₂H₄) ppm generation/month	<8	>7 <23	>22 <38	>37
Acetylene (C₂H₂) ppm	<36	>35 <51	>50 <81	>80
Acetylene (C₂H₂) ppm generation/month	<.5	>.4 <1.5	>1.49 <2.5	>2.49
Carbon Monoxide (CO) ppm	<351	>350 <571	>570 <1,401	>1,400
Carbon Monoxide (CO) ppm generation/month	<70	>69 <220	>219 <350	>349
Carbon Dioxide (CO₂) ppm	<2501	>2500 <4,001	>4,000 <10,001	>10,000
Carbon Dioxide (CO₂) ppm generation/month	<700	>699 <2,100	>2,099 <3,500	>3,499
CO₂/CO Ratio	>10	<10.1 >6	<6.1 >3.9	<4
Oxygen (O₂) ppm	<3,501	>3,500 <7,001	>7,000 <10,001	>10,000
Total Combustible Gas ppm	<721	>720 <1,921	>1,920 <4,631	>4,631
Moisture in Oil ppm **> 230 kv**	<6	>5 <9	>8 <13	>12
Moisture in Oil ppm **< = 230 kv**	<10	>9 <16	>15 <21	>20
Interfacial Tension (dynes per cm)	>45	<46 >35	<36 >25	<26
Acid Number (KOH in milligrams)	<.05	>.04 <.2	>.19 <.51	>.5
Dielectric Breakdown Voltage kV **>288 kV**	>55	>40 <56	>24 <41	<25
Dielectric Breakdown Voltage kV <288 kV	>55	>40 <56	>19 <41	<20
Power Factor at 25 °C %	<0.5	>0.49 <1.0	>0.99 <1.51	>1.5
Oxidation Inhibitor %	>0.3	>0.31 >0.2	>0.21 >0.09	<0.09
2FAL furans for 65 °C rise transformers	<150	>149 <200	>199 <251	>250
2FAL furans for 55 °C rise transformers	<450	>449 <600	>559 <751	>750

Table 7 – 345-kV Transformer Example

Tests	Value	Condition Status			
		1	2	3	4
		Good	Fair	Poor	Action
Hydrogen (H_2) ppm	1	TRUE	FALSE	FALSE	FALSE
Hydrogen (H_2) ppm generation/month	0.08	TRUE	FALSE	FALSE	FALSE
Methane (CH_4) ppm	3	TRUE	FALSE	FALSE	FALSE
Methane (CH_4) ppm gerneration/month	0.25	TRUE	FALSE	FALSE	FALSE
Ethane (C_2H_6) ppm	14	TRUE	FALSE	FALSE	FALSE
Ethane (C_2H_6) ppm generation/month	1.17	TRUE	FALSE	FALSE	FALSE
Ethylene (C_2H_4) ppm	1	TRUE	FALSE	FALSE	FALSE
Ethylene (C_2H_4) ppm generation/month	0.08	TRUE	FALSE	FALSE	FALSE
Acetylene (C_2H_2) ppm	0	TRUE	FALSE	FALSE	FALSE
Acetylene (C_2H_2) ppm generation/month	0	TRUE	FALSE	FALSE	FALSE
Carbon Monoxide (CO) ppm	6	TRUE	FALSE	FALSE	FALSE
Carbon Monoxide (CO) ppm generation/month	0.50	TRUE	FALSE	FALSE	FALSE
Carbon Dioxide (CO_2) ppm	493	TRUE	FALSE	FALSE	FALSE
Carbon Dioxide (CO_2) ppm generation/month	41.08	TRUE	FALSE	FALSE	FALSE
CO_2/CO Ratio	82.17	TRUE	FALSE	FALSE	FALSE
Oxygen (O_2) ppm	1,569	TRUE	FALSE	FALSE	FALSE
Total Combustible Gas ppm	25	TRUE	FALSE	FALSE	FALSE
Moisture in Oil ppm	5	TRUE	FALSE	FALSE	FALSE
Interfacial Tension (dynes per cm)	42.8	FALSE	TRUE	FALSE	FALSE
Acid Number (KOH in milligrams)	0.010	TRUE	FALSE	FALSE	FALSE
Dielectric Breakdown Voltage kV	34	FALSE	FALSE	TRUE	FALSE
Power Factor at 25 °C %	0.05	TRUE	FALSE	FALSE	FALSE
Oxidation Inhibitor %	0.05	FALSE	FALSE	FALSE	TRUE
2FAL furans for 55 °C rise transformer	200	TRUE	FALSE	FALSE	FALSE

CAUTION:

Transformer engineering expertise is needed in the use of this table. The action required may be as simple as sending an additional oil sample to the lab or as complex as conducting extensive testing on the transformer. It may also include repairs, internal inspection, and/or complete replacement of the transformer.

The condition ranges shown represent a composite of IEEE C57-104, IEC 60599, Delta X Research's Transformer Oil Analysis, and many years of transformer experience.

The table above was compiled by Clarence Herron of the Glen Canyon Field Office of Reclamation.

6.1.2 Key Gas Method

The key gas method of interpreting DGA is set forth in IEEE [12]. Key gases formed by degradation of oil and paper insulation are hydrogen (H_2), methane (CH_4), ethane (C_2H_6), ethylene (C_2H_4), acetylene (C_2H_2), carbon monoxide (CO), and oxygen (O_2). Except for carbon monoxide and oxygen, all these gases are formed from the degradation of the oil itself. Carbon monoxide, carbon dioxide (CO_2), and oxygen are formed from degradation of cellulose (paper) insulation. Carbon dioxide, oxygen, nitrogen (N_2), and moisture can also be absorbed from the air, if there is an oil/air interface or if there is a leak in the tank. Some of our transformers have a pressurized nitrogen blanket above the oil and, in these cases, nitrogen may be near saturation (see table 8). Gas type and amounts are determined by where the fault occurs in the transformer and the severity and energy of the event. Events range from low energy events, such as partial discharge, producing hydrogen and trace amounts of methane and ethane, to very high energy sustained arcing, capable of generating all the gases, including acetylene, requiring the most energy.

6.1.2.1 Four-Condition DGA Guide (IEEE C57-104) –
A four-condition DGA guide to classify risks to transformers with no

Table 8 – Dissolved Key Gas Concentration Limits (ppm)

Status	H_2	CH_4	C_2H_2	C_2H_4
Condition 1	100	120	35	50
Condition 2	101-700	121-400	36-50	51-100
Condition 3	701-1,800	401-1,000	51-80	101-200
Condition 4	>1,800	>1,000	>80	>200
Status	C_2H_6	CO	CO_2[1]	TDCG[2]
Condition 1	65	350	2,500	720
Condition 2	66-100	351-570	2,500-4,000	721-1,920
Condition 3	101-150	571-1,400	4,001-10,000	1,921-4,630
Condition 4	>150	>1,400	>10,000	>4,630

CAUTION:

Transformers generate some combustible gases from normal operation, and condition numbers for dissolved gases given in IEEE C57-104-1991™ [12] (table 8 above) are extremely conservative. Transformers can operate safely with individual gases in Condition 4 with no problems, provided they are stable and gases are not increasing or are increasing very slowly. If TDCG and individual gases are significantly increasing (more than 30 ppm per day [ppm/day]), an active fault is in progress. The transformer should be de-energized when Condition 4 levels are reached.

previous problems has been published in the Standard IEEE Standard (Std) [20], C57-104™. The guide uses combinations of individual gases and total combustible gas concentration as indicators. It is not universally accepted and is only one of the tools used to evaluate dissolved gas in transformers. The four IEEE® conditions are defined below, and gas levels are in table 8.

Condition 1: Total dissolved combustible gas (TDCG) below this level indicates the transformer is operating satisfactorily.

Any individual combustible gas exceeding specified levels in table 8 should have additional investigation.

Condition 2: TDCG within this range indicates greater than normal combustible gas level. Any individual combustible gas exceeding specified levels in table 8 should have additional investigation. A fault may be present. Take DGA samples at least often enough to calculate the amount of gas generation per day for each gas. (See table 9 for recommended sampling frequency and actions.)

Condition 3: TDCG within this range indicates a high level of decomposition of cellulose insulation and/or oil. Any individual combustible gas exceeding specified levels in table 6 should have additional investigation. A fault or faults are probably present. Take DGA samples at least often enough to calculate the amount of gas generation per day for each gas (see table 9).

Condition 4: TDCG within this range indicates excessive decomposition of cellulose insulation and/or oil. Continued operation could result in failure of the transformer (see table 9).

A sudden increase in key gases and the rate of gas production is more important in evaluating a transformer than the accumulated amount of gas. One very important consideration is acetylene (C_2H_2). Generation of any amount of this gas above a few ppm indicates high-energy arcing. Trace amounts (a few ppm) can be generated by a very hot thermal fault (500 C or higher). A one-time arc, caused by a nearby lightning strike or a high-voltage surge, can also generate a small amount of C_2H_2. If C_2H_2 is found in the DGA, oil samples should be taken weekly, or even daily, to determine if additional C_2H_2 is being generated. If no additional acetylene is found and the level is below the IEEE® Condition 4, the transformer may continue in service. However, if acetylene continues to increase, the transformer has an active high-energy internal arc and should be taken out of service immediately. Further operation is extremely

Table 9 – Actions Based on Dissolved Combustible Gas (Adapted from [12])

Conditions	TDCG Level or Highest Individual Gas (See table 8)	TDCG Generation Rates (ppm/day)	Sampling Intervals and Operating Actions for Gas Generation Rates	
			Sampling Interval	Operating Procedures
Condition 1	<720 ppm of TDCG or highest condition based on individual combustible gas from table 8.	<10	Annually: 6 months for extra high-voltage transformer	Continue normal operation.
		10-30	Quarterly	
		>30	Monthly	Exercise caution. Analyze individual gases to find cause. Determine load dependence.
Condition 2	721–1,920 ppm of TDCG or highest condition based on individual combustible gas from table 8.	<10	Quarterly	Exercise caution. Analyze individual gases to find cause. Determine load dependence.
		10-30	Monthly	
		>30	Monthly	
Condition 3	1,941–2,630 ppm of TDCG or highest condition based on individual combustible gas from table 8.	<10	Monthly	Exercise extreme caution. Analyze individual gases to find cause. Plan outage. Call manufacturer and other consultants for advice.
		10-30	Weekly	
		>30	Weekly	
Condition 4	>4,639 ppm of TDCG or highest condition based on individual combustible gas from table 8.	<10	Weekly	Exercise extreme caution. Analyze individual gases to find cause. Plan outage. Call manufacturer and other consultants for advice.
		10-30	Daily	
		>30	Daily	Consider removal from service. Call manufacturer and other consultants for advice.

NOTES:

1. Either the highest condition based on individual combustible gas or TDCG can determine the condition (1, 2, 3, or 4) of the transformer. For example, if the TDCG is between 1,941 ppm and 2,630 ppm, this indicates Condition 3. However, if hydrogen is greater than 1,800 ppm, the transformer is in Condition 4, as shown in table 8.

2. When the table says "determine load dependence," this means to try to find out if the gas generation rate in ppm/day goes up and down with the load. The transformer may be overloaded or have a cooling problem. Take oil samples every time the load changes; if load changes are too frequent, this may not be possible.

3. To get the TDCG generation rate, divide the change in TDCG by the number of days between samples that the transformer has been loaded. Down days should not be included. The individual gas generation rate in ppm/day is determined by the same method.

hazardous and may result in explosive catastrophic failure of the tank, spreading flaming oil over a large area.

Table 9 assumes that no previous DGA tests were performed on the transformer or that no recent history exists. If a previous DGA exists, it should be reviewed to determine if the situation is stable (gases are not increasing significantly) or unstable (gases are increasing significantly).

Before going to table 11, determine the transformer status from table 10; that is, look at the DGA and see if the transformer is in Condition 1, 2, 3, or 4. The condition for a particular transformer is determined by finding the highest level for any **individual gas** or by using the TDCG [12]. Either the individual gas or the TDCG can give the transformer a higher condition number, which means it is at greater risk. If the TDCG number shows the transformer in Condition 3 and an individual gas shows the transformer in Condition 4, the transformer is in Condition 4. Always be conservative and assume the worst until proven otherwise.

6.1.3 Sampling Intervals and Recommended Actions

When sudden increases occur in dissolved gases, the procedures recommended in table 9 should be followed. Table 9 is **paraphrased** from table 3 in IEEE C57.104-1991 [12]. To make it easier to read, the order has been reversed with Condition 1 (lowest risk transformer) at the top and Condition 4 (highest risk) at the bottom. The table indicates the recommended sampling intervals and actions for various levels of TDCG in ppm.

An increasing gas generation rate indicates a problem of *increasing severity*; therefore, as the generation rate (ppm/day) increases, a shorter sampling interval is recommended (see table 9).

Some information has been added to the table from IEEE C57-104-1991 (inferred from the text). To see the exact table, refer to the IEEE Standard [12].

If the cause of gassing can be determined and the risk can be assessed, the sampling interval may be extended. For example, if the core is meggered and an additional core ground is found, even though table 9 may recommend a monthly sampling interval, an operator may choose to lengthen the sampling interval, because the source of the gassing and generation rate is known.

A decision should never be made on the basis of just one DGA. It is very easy to contaminate the sample by accidentally exposing it to air. Mislabeling a sample is also a common cause of error. Mislabeling could occur when the sample is taken, or it could be accidentally contaminated or mishandled at the laboratory. Mishandling may allow some gases to escape to the atmosphere and other gases, such as oxygen, nitrogen, and carbon dioxide, to migrate from the atmosphere into the sample. **If you notice a transformer problem from the DGA, the first thing to do is take another sample for comparison.**

The gas generation chart (figure 48) [14, 17] and the discussion below represent only an approximate temperature at which gases form. The figure is not drawn to scale and is only for purposes of illustrating temperature relationships, gas types, and quantities. These relationships represent what generally has been proven in controlled laboratory conditions using a mass spectrometer. This chart was used by R.R. Rogers of the Central Electric Generating Board (CEGB) of England to develop the "Rogers Ratio Method" of analyzing transformers (discussed in section 6.1.9.4).

A vertical band on the left of the chart shows what gases and approximate relative quantities are produced under partial discharge conditions. Note that all the gases are given off, but in much less quantity than hydrogen. It takes only a very low energy event (partial discharge/corona) to cause hydrogen molecules to form from the oil.

Gases are formed inside an oil-filled transformer similar to a petroleum refinery still, where various gases begin forming at specific temperatures. From the Gas Generation Chart, we can see relative

COMBUSTIBLE GAS GENERATION VS.
APPROXIMATE OIL DECOMPOSITION TEMPERATURE

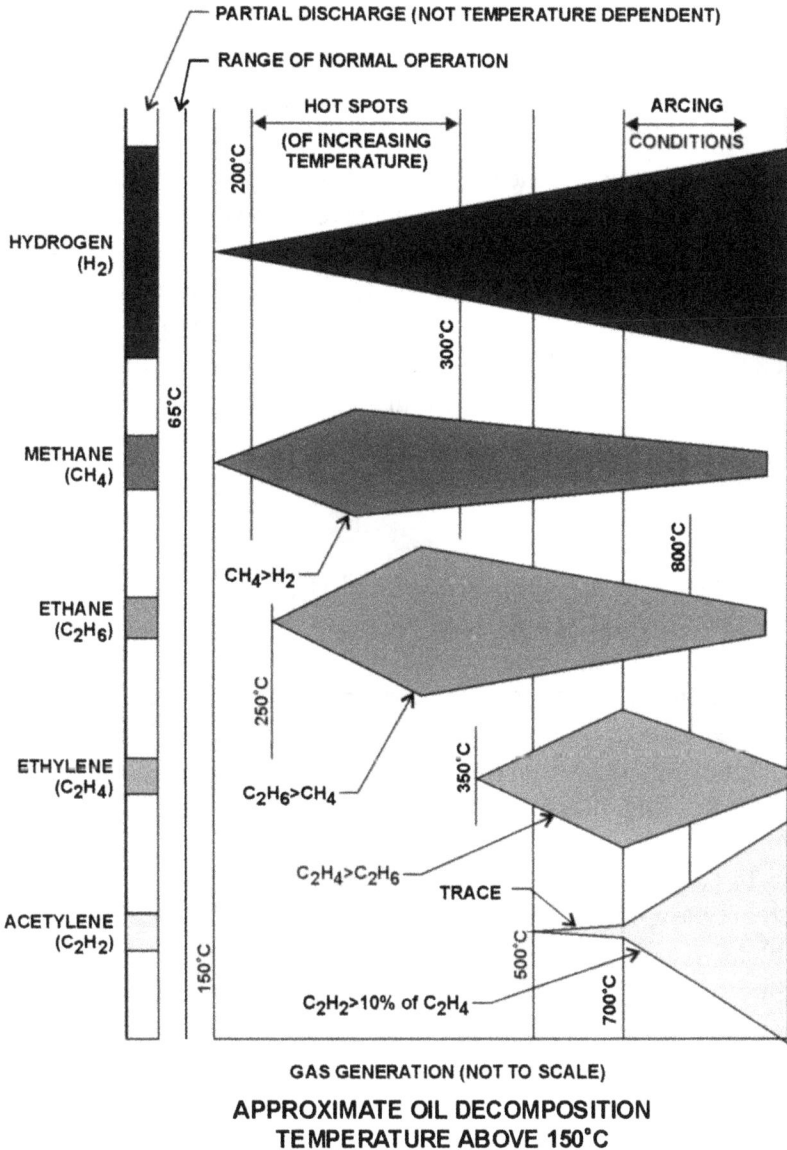

PARTIAL DISCHARGE (NOT TEMPERATURE DEPENDENT)

RANGE OF NORMAL OPERATION

HOT SPOTS

(OF INCREASING TEMPERATURE)

ARCING

CONDITIONS

HYDROGEN (H_2)

200°C

300°C

METHANE (CH_4)

65°C

$CH_4 > H_2$

800°C

ETHANE (C_2H_6)

250°C

ETHYLENE (C_2H_4)

$C_2H_6 > CH_4$

350°C

$C_2H_4 > C_2H_6$

TRACE

ACETYLENE (C_2H_2)

150°C

500°C

$C_2H_2 > 10\%$ of C_2H_4

700°C

GAS GENERATION (NOT TO SCALE)

APPROXIMATE OIL DECOMPOSITION
TEMPERATURE ABOVE 150°C

Figure 48 – Combustible Gas Generation Versus Temperature.

amounts of gas as well as approximate temperatures. Hydrogen and methane begin to form in small amounts around 150 °C. Notice from the chart that beyond maximum points, methane (CH_4), ethane and ethylene production goes down as temperature increases. At about 250 °C, production of ethane (C_2H_6) starts. At about 350 °C, production of ethylene (C_2H_4) begins. Acetylene (C_2H_2) production begins between 500 °C and 700 °C. In the past, the presence of only trace amounts of acetylene (C_2H_2) was considered to indicate that a temperature of at least 700 °C had occurred; however, recent discoveries have led to the conclusion that a thermal fault (hot spot) of 500 °C can produce trace amounts (a few ppm). Larger amounts of acetylene can only be produced above 700 °C by internal arcing. Notice that between 200 °C and 300 °C, the production of methane exceeds hydrogen. Starting at about 275 °C and on up, the production of ethane exceeds methane. At about 450 °C, hydrogen production exceeds all others until about 750 °C to 800 °C; then more acetylene is produced.

It should be noted that small amounts of H_2, CH_4, and CO are produced by normal aging. Thermal decomposition of oil-impregnated cellulose produces CO, CO_2, H_2, CH_4, and O_2. Decomposition of cellulose insulation begins at only about 100 °C or less. Therefore, operation of transformers at no more than 90 °C is imperative. Faults will produce internal "hot spots" of far higher temperatures than these, and the resultant gases show up in the DGA.

Table 11 (later in this section) is a chart of "fault types," parts of which are paraphrased from the IEC 60599 [13]. This chart is not complete. It is impossible to chart every cause and effect due to the extreme complexity of transformers. DGAs must be carefully examined with the idea of determining possible faults and possible courses of action. These decisions are based on judgment and experience and are seldom "cut and dried." Most professional associations agree that there are two basic fault types, thermal and electrical. The first three on the chart are electrical discharges, and the last three are thermal faults.

Ethane and ethylene are sometimes called "hot metal gases." When these gases are generated and acetylene is not, the problem found inside the transformer normally involves hot metal. This may include bad contacts on the tap changer or a bad connection somewhere in the circuit, such as a main transformer lead. Stray flux impinging on the tank (such as in Westinghouse 7M series transformers) can cause these "hot metal gases." Sometimes, a shield becomes loose and falls and becomes ungrounded. Static can then build up and discharge to a grounded surface, producing "hot metal" gases. An unintentional core ground with circulating currents can also produce these gases. There are many other examples.

Notice that both types of faults (thermal and electrical) may be occurring at the same time, and one may cause the other. The associations do not mention magnetic faults; however, magnetic faults (such as stray magnetic flux impinging the steel tank or other magnetic structures) also cause hot spots.

6.1.4 Atmospheric Gases

Atmospheric gases (N_2, CO_2, and O_2) can be very valuable in a DGA in revealing a possible leak. However, as mentioned elsewhere, there are other reasons these gases are found in DGAs. Nitrogen may exist from shipping the transformer with N_2 inside or from a nitrogen blanket. CO_2 and O_2 are formed by degradation of cellulose. Be very careful; look at several DGAs and see if atmospheric gases and, possibly, moisture levels are increasing. Also, carefully look at the transformer to see if you can find an oil leak. Moisture and atmospheric gases will leak inside when the transformer is off and the ambient temperature drops (see section 6.1.11 on moisture).

6.1.5 Dissolved Gas Software

Several companies offer DGA computer software that diagnoses transformer problems. These diagnoses must be used with engineering judgment and should never be taken at face value. The software is constantly changing. The Technical Service Center uses "Transformer Oil Analyst" (TOA) by Delta X Research that uses a composite of several current DGA methods. Dissolved gas analysis help is available from the TSC through the D-8440 and D-8450 groups. Both groups have the above software and experience in diagnosing transformer problems.

One set of rules that TOA uses to generate alarms is based loosely on IEC 60599 (table 10). These rules are also very useful in daily dissolved gas analysis, which are based on L1 limits of IEC 60599, except for acetylene. IEC 60599 gives a range for L1 limits instead of a specific value. TOA uses the average in this range and then gives the user a "heads up" if a generation rate exceeds 10% of L1 limits per month. Acetylene is the exception; IEEE sets an L1 limit of 35 ppm (too high), and IEC sets an acetylene range of 3 to 50. TOA picks the

Table 10 – TOA L1 Limits and Generation Rate Per Month Alarm Limits

Gas	L1 Limits	G1 Limits (ppm per month)	G2 Limits (ppm per month)
H_2	100	10	50
CH_4	75	8	38
C_2H_2	3	3	3
C_2H_4	75	8	38
C_2H_6	75	8	38
CO	700	70	350
CO_2	7,000	700	3,500

NOTE:

If one or more gas generation rates are equal to or exceed G1 limits (10% of L1 limits per month), you should begin to pay more attention to this transformer. Reduce the DGA sample interval, reduce loading, plan for future outage, contact the manufacturer, etc.

lowest number (3 ppm) and sets the generation rate alarm value at 3 ppm per month. L1 limits are the quantities at which you should begin to watch the transformer more closely (i.e., the first level of concern).

If one or more combustible gas generation rates are equal to or exceed G2 limits (50% of L1 limits per month), this transformer should be considered in critical condition. You may want to reduce sample intervals to monthly or weekly, plan an outage, plan to rebuild or replace the transformer, etc. If an active arc is present (C_2H_2 generation), or if other heat gases are high (above Condition 4 limits in table 8), and G2 limits are exceeded, the transformer should be de-energized.

Table 11 is taken from IEC 60599 and contains possible faults and possible findings. This chart is not all inclusive and should be used with other information. Additional possible faults are listed on the following and preceding pages.

Transformers are so complex that it is impossible to put all symptoms and causes into a chart. Several additional transformer problems are listed below; any of these may generate gases:

1. Gases are generated by normal operation and aging, mostly H_2 and CO, with some CH_4. H_2 is the easiest gas to produce except possibly CO. Production of H_2 and other gases can be caused by partial discharge (corona), sharp corners on bottom bushing connectors, loose core ground, wet spot on core from gasket leak above, loose corona shield on bottom of bushing, loose tap changer shield, etc. H_2 is not very stable when dissolved in oil. Consecutive DGAs may show variation in amounts of H_2 and other unstable gases. Acetylene is the most stable gas; variation in amounts of this gas in the upward direction means the transformer has an active arcing fault. If the variation is going up and down within detection

Table 11 – Fault Types

Key Gases	Possible Faults	Possible Findings
H_2, possible trace of CH_4 and C_2H_6. Possible CO.	Partial discharges (corona).	Weakened insulation from aging and electrical stress.
H_2, CH_4, (some CO if discharges involve paper insulation). Possible trace amounts of C_2H_6.	Low energy discharges (sparking). (May be static discharges).	Pinhole punctures in paper insulation with carbon and carbon tracking. Possible carbon particles in oil. Possible loose shield, poor grounding of metal objects
H_2, CH_4, C_2H_6, C_2H_4, and the key gas for arcing C_2 H_2 will be present perhaps in large amounts. If $C_2 H_2$ is being generated, arcing is continuing. CO will be present if paper is being heated.	High energy discharges (arcing).	Metal fusion (poor contacts in tap changer or lead connections). Weakened insulation from aging and electrical stress. Carbonized oil. Paper destruction if it is in the arc path or overheated.
H_2, CO.	Thermal fault less than 300 °C in an area close to paper insulation (paper is being heated).	Discoloration of paper insulation. Overloading and/or cooling problem. Bad connection in leads or tap changer. Stray current path and/or stray magnetic flux.
H_2, CO, CH_4, C_2H_6, $C_2 H_4$.	Thermal fault between 300 °C and 700 °C.	Paper insulation destroyed. Oil heavily carbonized.
All the above gases and acetylene in large amounts.	High energy electrical arcing 700 °C and above.	Same as above with metal discoloration. Arcing may have caused a thermal fault.

limits of the test equipment (section 6.1.9.4, table 14) in consecutive DGAs, this is simply a variation of the lab's test equipment and personnel.

2. Operating transformers at sustained overload will generate combustible gases.

3. Problems with cooling systems, discussed in sections 3.3.1 and 3.4.5, can cause overheating.

4. A blocked oil duct inside the transformer can cause local overheating, generating gases.

5. An oil directing baffle loose inside the transformer causes misdirection of cooling oil.

6. Oil circulating pump problems (bearing wear, impeller loose or worn, or pump running in reverse) can cause transformer cooling problems.

7. Oil level is too low; this will not be obvious if the level indicator is inoperative.

8. Sludge in the transformer and cooling system (see section 3.4.5.4).

9. Circulating stray currents may occur in the core, structure, and/or tank.

10. An unintentional core ground may cause heating by providing a path for stray currents.

11. A hot-spot can be caused by a bad connection in the leads or by a poor contact in the tap changer.

12. A hot-spot may also be caused by discharges of static electrical charges that build up on shields or core and structures and that are not properly grounded.

13. Hot-spots may be caused by electrical arcing between windings and ground, between windings of different potential, or in areas of different potential on the same winding, due to deteriorated or damaged insulation.

14. Windings and insulation can be damaged by faults downstream (through faults), causing large current surges through the windings. Through faults cause extreme magnetic and physical forces that can distort and loosen windings and wedges. The results may be arcing in the transformer, beginning at the time of the fault, or the insulation may be weakened and arcing will develop later.

15. Insulation can also be damaged by a voltage surge, such as a nearby lightning strike, switching surge, or closing out of step, which may result in immediate arcing or arcing that develops later.

16.   Insulation may be deteriorated from age and worn out. Clearances and dielectric strength are reduced, allowing partial discharges and arcing to develop. This can also reduce physical strength, allowing wedging and windings to move extensively during a through-fault, causing total mechanical and electrical failure.

17.   High noise level (hum due to loose windings or core laminations) can generate gas due to heat from friction. Compare the noise to sister transformers, if possible. Sound level meters are available at the TSC for diagnostic comparison and to establish baseline noise levels for future comparison.

6.1.6 Temperature

Gas production rates increase exponentially with temperature and directly with volume of oil and paper insulation at high enough temperature to produce gases [12]. Temperature decreases as distance from the fault increases. Temperature at the fault center is highest, and oil and paper, here, will produce the most gas. As distance increases from the fault (hot spot), temperature goes down, and the rate of gas generation also goes down. Because of the volume effect, a large heated volume of oil and paper will produce the same amount of gas as a smaller volume at a higher temperature [12]. We cannot tell the difference by looking at the DGA. This is one reason that interpreting DGAs is not an exact science.

6.1.7 Gas Mixing

Concentration of gases in close proximity to an active fault will be higher than in the DGA oil sample. As distance increases from a fault, gas concentrations decrease. Equal mixing of dissolved gases in the total volume of oil depends on time and oil circulation. If there are no pumps to force oil through radiators, complete mixing of gases in the total oil volume takes longer. With pumping and normal loading, complete mixing equilibrium should be reached within

24 hours and will have little effect on DGA if an oil sample is taken 24 hours or more after a problem begins.

6.1.8 Gas Solubility

Solubilities of gases in oil vary with temperature and pressure [14]. Solubility of all transformer gases vary proportionally up and down with pressure. Variation of solubilities with temperature is much more complex. Solubilities of hydrogen, nitrogen, carbon monoxide, and oxygen go up and down proportionally with temperature. Solubilities of carbon dioxide, acetylene, ethylene, and ethane are reversed and vary inversely with temperature changes. As temperature rises, solubilities of these gases go down; and as temperature falls, their solubilities increase. Methane solubility remains almost constant with temperature changes. Table 12 is accurate only at standard temperature and pressure (STP), (0 °C/32 °F) and (14.7 psi/29.93 inches of mercury, which is standard barometric pressure at sea level). Table 12 shows only relative differences in how gases dissolve in transformer oil.

From the solubility table (table 12) below, comparing hydrogen with a solubility of 7% and acetylene with a solubility of 400%, you can see that transformer oil has a much greater capacity for dissolving acetylene. However, 7% hydrogen by volume represents 70,000 ppm, and 400% acetylene represents 4,000,000 ppm. You probably will never see a DGA with numbers this high. Nitrogen can approach maximum level if there is a pressurized nitrogen blanket above the oil. Table 12 shows the maximum amount of each gas that the oil is capable of dissolving at standard temperature and pressure. At these levels, the oil is said to be saturated.

If you have conservator-type transformers and nitrogen, oxygen, and CO_2 are increasing, there is a good possibility that the tank has a leak or that the oil may have been poorly processed. Check the diaphragm or bladder for leaks (section 4.9) and check for oily residue around the pressure relief device and other gasketed openings. There should be

Table 12 – Dissolved Gas Solubility in Transformer Oil Accurate Only at STP, 0 °C (32 °F) and 14.7 psi (29.93 Inches of Mercury)

Dissolved Gas	Formula	Solubility in Transformer Oil (% by Volume)	Maximum Equivalent (ppm by Volume)	Primary Causes/ Sources
Hydrogen[1]	H_2	7.0	70,000	Partial discharge, corona, electrolysis of H_2O
Nitrogen	N_2	8.6	86,000	Inert gas blanket, atmosphere
Carbon Monoxide[1]	CO	9.0	90,000	Overheated cellulose, air pollution
Oxygen	O_2	16.0	160,000	Atmosphere
Methane[1]	CH_4	30.0	300,000	Overheated oil
Carbon Dioxide	CO_2	120.0	1,200,000	Overheated cellulose, atmosphere
Ethane[1]	C_2H_6	280.0	2,800,000	Overheated oil
Ethylene[1]	C_2H_4	280.0	2,800,000	Very overheated oil
Acetylene[1]	C_2H_2	400.0	4,000,000	Arcing in oil

[1] Denotes combustible gas. Overheating can be caused both by high temperatures and by unusual or abnormal electrical stress.

fairly low nitrogen and especially low oxygen in a conservator-type transformer. However, if the transformer was shipped new with pressurized nitrogen inside and has not been de-gassed properly, there may be high nitrogen content in the DGA, but the nitrogen level should not be increasing after the transformer has been in service for a few years. When oil is installed in a new transformer, a vacuum is placed on the tank that pulls out nitrogen and pulls in the oil. Oil is free to absorb nitrogen at the oil/gas interface, and some nitrogen may be trapped in the windings, paper insulation, and structure. In this case, nitrogen may be fairly high in the DGAs. However, oxygen should be very low, and nitrogen should not be increasing. It is important to take an oil sample early in the transformer's service life to establish a baseline DGA; then take samples at least annually. The nitrogen and oxygen can be compared with earlier DGAs. If they increase, it is a good indication of a leak. If the transformer oil has ever been de-gassed, nitrogen and oxygen should be low in the DGA. It is extremely important to keep accurate records over a transformer's life; when a problem occurs, recorded information helps greatly in troubleshooting.

6.1.9 Diagnosing a Transformer Problem Using Dissolved Gas Analysis and the Duval Triangle

---

CAUTION:

Do not use the Duval Triangle (figure 49) to determine whether or not a transformer has a problem. Notice, there is no area on the triangle for a transformer that does not have a problem. The triangle will show a fault for every transformer whether it has a fault or not. Use the above IEEE method or table 13 to determine if a problem exists before applying the Duval Tirangle. The Duval Triangle is used only to determine what the problem is. As with other methods, a significant amount of gas (at least L1 limits and G2 generation rates in table 13) must already be present before this method is valid.

6.1.9.1 Origin of the Duval Triangle – Michel Duval of Hydro Quebec developed this method in the 1960s using a database of thousands of DGAs and transformer problem diagnosis. More recently, this method was incorporated in the Transformer Oil Analyst Software version 4 (TOA 4), developed by Delta X Research and used by many in the utility industry to diagnose transformer problems. This method has proven to be accurate and dependable over many years and is now gaining in popularity. The method and how to use it are described below.

6.1.9.2 How to Use the Duval Triangle –
1.   First determine whether a problem exists by using the IEEE® method above, and/or table 13 below. At least one of the hydrocarbon gases or hydrogen must be in

PD = PARTIAL DISCHARGE
T1 = THERMAL FAULT LESS THAN 300°C
T2 = THERMAL FAULT BETWEEN
      300°C AND 700°C
T3 = THERMAL FAULT GREATER THAN 700°C
D1 = LOW ENERGY DISCHARGE (SPARKING)
D2 = HIGH ENERGY DISCHARGE (ARCHING)
DT = MIX OF THERMAL AND
      ELECTRICAL FAULTS

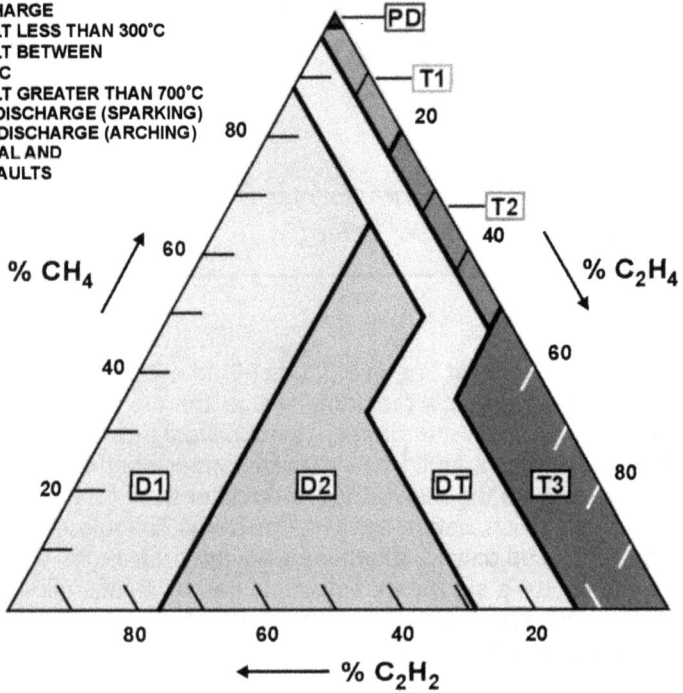

Figure 49 – The Duval Triangle.

Table 13 – L1 Limits and Generation Rate Per Month Limits

Gas	L1 Limits	G1 Limits (ppm per month)	G2 Limits (ppm per month)
H_2	100	10	50
CH_4	75	8	38
C_2H_2	3	3	3
C_2H_4	75	8	38
C_2H_6	75	8	38
CO	700	70	350
CO_2	7,000	700	3,500

IEEE® Condition 3 and increasing at a generation rate (G2), from table 13, before a problem is confirmed. To use table 13 without the IEEE® method, at least one of the individual gases must be at L1 level or above and the gas generation rate must be at least at G2.

The L1 limits and gas generation rates from table 13 are more reliable than the IEEE® method; however, one should use both methods to confirm that a problem exists. If there is a sudden increase in H_2 with only carbon monoxide and carbon dioxide and little or none of the hydrocarbon gases, use section 6.1.10 (CO_2/CO ratio) to determine if the cellulose insulation is being degraded by overheating.

2. Once a problem has been determined to exist, use the total accumulated amount of the three Duval Triangle gases and plot the percentages of the total on the triangle to arrive at a diagnosis. An example is shown below. Also, calculate the amount of the three gases used in the Duval Triangle, generated since the sudden increase in gas began. Subtracting out the amount of gas generated prior to the sudden increase will give the amount of gases generated since the fault began. Detailed instructions and an example are shown below.

    a. Take the amount (ppm) of methane in the DGA and subtract the amount of CH_4 from an earlier DGA, before the sudden increase in gas. This will give the amount of methane generated since the problem started.

    b. Repeat this process for the remaining two gases, ethylene and acetylene.

3. Add the three numbers (differences) obtained by the process of step 2 above. This gives 100 % of the three key gases, generated since the fault, used in the Duval Triangle.

4. Divide each individual gas difference by the total difference of gas obtained in step 3 above. This gives the percentage of increase of each gas of the total increase.

5.   Plot the percentage of each gas on the Duval Triangle,
     beginning on the side indicated for that particular gas. Draw
     lines across the triangle for each gas, parallel to the hash
     marks shown on each side of the triangle. An example is
     shown below.

NOTE:

In most cases, acetylene will be zero, and the result will be a
point on the right side of the Duval Triangle.

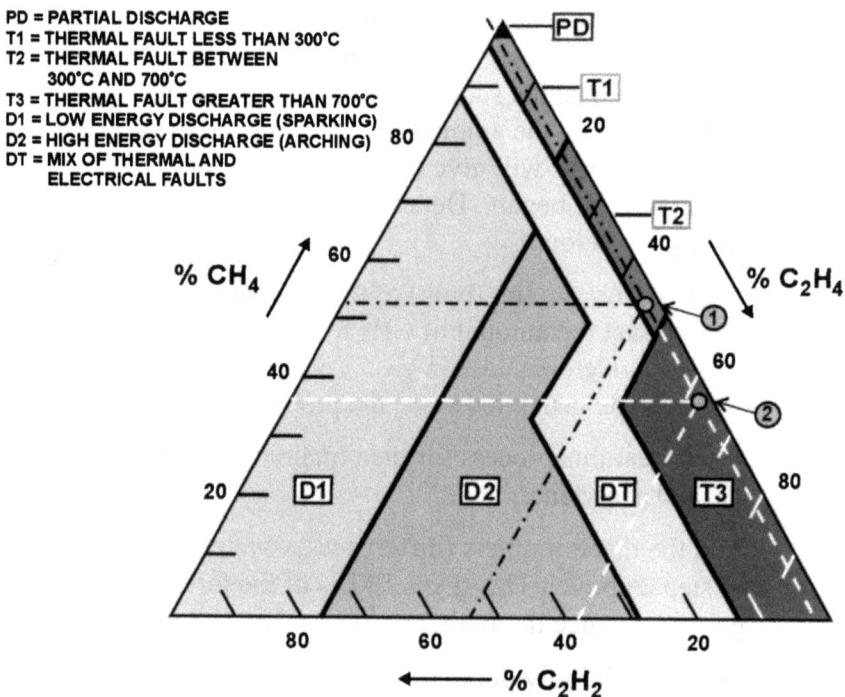

Figure 50 – Duval Triangle Diagnostic Example of a Reclamation Transformer.

Compare the total accumulated gas diagnosis and the diagnosis obtained by using only the increase-in-gases after a fault. If the fault has existed for some time, or if generation rates are high, the two diagnoses will be the same. If the diagnoses are not the same, always use the diagnosis of the increase in gases generated by the fault, which will be the more severe of the two. See the example below of a Reclamation transformer where the diagnosis using the increase in gas is more severe than when using the total accumulated gas.

> **Example:** Using figure 50 and the information below, two diagnoses of a Reclamation transformer were obtained. The first diagnosis (Point 1) was obtained using the total amount of the three gases used by the Duval Triangle. The second diagnosis (Point 2) was obtained using only the increase in gases between the two DGAs. CO and CO_2 are used to evaluate cellulose.

	DGA No. 1	DGA No. 2	Increase		DGA No. 1	DGA No. 2	Increase
CH_4	142	192	50	CO	176	199	23
C_2H_4	84	170	86	CO_2	1,009	2,326	1,317
C_2H_2	4	7	3				
Total	230	369	139				

Steps to Obtain the First Diagnosis (Point 1) on the Duval Triangle (Figure 50)

1. Use the **total accumulated** gas from DGA 2 = 369

2. Divide each gas by the total to find the percentage of each gas of the total.

   % CH_4 = 192/369 = 52%, % C_2H_4 = 170/369 = 46%,
   % C_2H_2 = 7/369 = 2%

3. Draw three lines across the Duval Triangle starting at the percentages obtained in step 2. These lines **must be drawn parallel** to the hash mark on each respective side. See the black dashed lines in figure 50 above.

4. Point 1 is obtained where the lines intersect within the T2 diagnostic area of the triangle, which indicates a thermal fault between 300 and 700 °C. See figure 49, Legend, above.

Steps to Obtain the Second Diagnosis (Point 2) on the Duval Triangle (Figure 50)

1. Use the **total increase** in gas = 139.

2. Divide each gas increase by the total increase to find the percentage of each gas of the total:

   % increase $CH_4 = 50/139 = 36\%$

   % increase $C_2H_4 = 86/139 = 46\%$
   % increase $C_2H_2 = 3/139 = 2\%$

3. Draw three lines across the Duval Triangle starting at the percentages obtained in step 2. These lines **must be drawn parallel** to the hash mark on each respective side. See the white dashed lines in figure 50 above. Note that C_2H_2 was the same percentage (2%) both times; and, therefore, both lines are the same.

4. Point 2 is obtained where the lines intersect within the T3 diagnostic area of the triangle, which indicates a thermal fault greater than 700 °C. See figure 49, Legend, above.

The ratio of total accumulated gas is $CO_2/CO = 2,326/199 = 11.7$.

The ratio of increase is $CO_2/CO = 1,317/23 = 57$. Neither of these ratios is low enough to cause concern. This shows that the thermal fault is not close enough to the cellulose insulation to cause heat degradation of the insulation. The large increase in CO_2 could mean an atmospheric leak.

NOTE:

1. Point 2 is the more severe diagnosis obtained by using the increase in gas rather than the total accumulated gas. It is helpful to perform both methods as a check; many times both diagnoses will come out the same.

2. CO and CO_2 are included to show that the fault does not involve severe degradation of cellulose insulation. See section 6.1.10 for an explanation of CO_2/CO ratios.

The fault is probably a bad connection on a bushing bottom, a bad contact or connection in the tap changer, or a problem with a core ground. These problems are probably all reparable in the field. Any of these problems can cause the results revealed by the Duval Triangle diagnosis above. These are areas where a fault will not degrade cellulose insulation, which would cause the CO_2/CO ratio to be much lower than what was obtained. For information to arrive at a probable fault see section 6.1.10.

6.1.9.3 Expertise Needed – A transformer expert should be consulted if a problematic trend is evidenced by a number of DGAs. The transformer manufacturer should be consulted along with DGA lab personnel and others experienced in transformer maintenance and diagnostics. Never make a diagnosis based on one DGA; a sample may have been mishandled or mislabeled, either in the field or lab.

6.1.9.4 Rogers Ratio Method of DGA – Rogers Ratio Method of DGA [19] is an additional tool that may be used to look at dissolved gases in transformer oil. Rogers Ratio Method compares quantities of different key gases by dividing one into the other. This gives a ratio of the amount of one key gas to another. By looking at the Gas Generation Chart (figure 48), you can see that, at certain temperatures, one gas will be generated more than another gas. Rogers used these relationships and determined that if a certain ratio existed, then a specific temperature had been reached. By comparing a large number of transformers with similar gas ratios and data collected when the transformers were examined, Rogers then could then say that certain faults were present. Like the key gas analysis above, this method is not a "sure thing" and is only an additional tool to use in analyzing transformer problems. Rogers Ratio Method, using three-key gas ratios, is based on earlier work by Doerneburg, who used five-key gas ratios. Ratio methods are only valid if a significant amount of the gases used in the ratio is present. A good rule is: **Never make a decision based only on a ratio if either of the two gases used in a ratio is less than 10 times the amount the gas chromatograph can detect [13].** This rule makes sure that instrument inaccuracies have little effect on the ratios. If either of the gases are lower than 10 times the detection limit, most likely you do not have the particular problem that this ratio represents. If the gases are not at least 10 times these limits does not mean you cannot use the Rogers Ratios; it means that the results are not as certain as if the gases were at least at these levels. This is another reminder that DGAs are not an exact science and there there is no "one best, easy way" to analyze transformer problems. Approximate detection limits are as follows, depending on the lab and equipment (table 14).

Table 14 – Dissolved Gas Analysis Detection Limits

Gas	Detection Limit
Hydrogen (H_2)	5 ppm
Methane (CH_4)	1 ppm
Acetylene (C_2H_2)	1 to 2 ppm
Ethylene (C_2H_2)	1 ppm
Ethane (C_2H_6)	1 ppm
Carbon monoxide (CO)	25 ppm
Carbon dioxide (CO_2)	25 ppm
Oxygen (O_2)	50 ppm
Nitrogen (N_2)	50 ppm

When a fault occurs inside a transformer, there is no problem with minimum gas amounts at which the ratio is valid. There will be more than enough gas present.

If a transformer has been operating normally for some time and a DGA shows a sudden increase in the amount of gas, first take a second sample to verify that there is a problem. If the next DGA shows gases to be more in line with prior DGAs, the earlier oil sample was contaminated, and there is no further cause for concern. If the second sample also shows increases in gases, the problem is real. To apply Ratio Methods, subtract gases that were present prior to sudden gas increases. This takes out gases that have been previously generated due to normal aging and from prior problems. This is especially true for ratios using H_2 and the cellulose insulation gases CO and CO_2 [13], which are generated by normal aging.

Rogers Ratio Method Uses the Following Three Ratios.

$$C_2H_2/C_2H_4, \quad CH_4/H_2, \quad C_2H_4/C_2H_6$$

These ratios and the resultant fault indications are based on large numbers of DGAs and transformer failures and what was discovered after the failures.

There are other ratio methods, but only the Rogers Ratio Method will be discussed since it is the one most commonly used. The method description is paraphrased from Rogers' original paper [19] and from IEC 60599 [13].

Ethylene and ethane are sometimes called "hot metal gases." Notice that this fault does not involve paper insulation, because CO is very low. H_2 and C_2H_2 are both less than 10 times the detection limit. This means the diagnosis does not have a 100% confidence level of being correct. However, due to the high ethylene, the fault is probably a bad connection where an incoming lead is bolted to a winding lead, perhaps bad tap changer contacts, or additional core ground (large

CAUTION:

Rogers Ratio Method is for **fault analyzing**, not for **fault detection**. You **must have already decided** that you have a problem from the total amount of gas (using IEEE limits) or increased gas generation rates. Rogers Ratios will only give you an indication of what the problem is; it cannot tell you whether or not you have a problem. If you already suspect a problem, based on total combustible gas levels or increased rate-of-generation, then you normally already will have enough gas for this method to work. A good system to determine whether you have a problem is to use table 9 in the Key Gas Method. If two or more of the key gases are in Condition 2 and the gas generation is at least 10% per month of the L1 limit, you have a problem. Also, for the diagnosis to be valid, gases used in ratios should be at least 10 times the detection limits given earlier. The more gas you have, the more likely the Rogers Ratio Method will give a valid diagnosis. The reverse is also true; the less gas you have, the less likely the diagnosis will be valid. If a gas used in the denominator of any ratio is zero, or is shown in the DGA as not detected, use the detection limit of that particular gas as the denominator. This gives a reasonable ratio to use in diagnostic table 15. Zero codes mean that you do not have a problem in this area.

Example 1
Example of a Reclamation Transformer DGA:

	Parts per Million
Hydrogen (H_2)	9
Methane (CH_4)	60
Ethane (C_2H_6)	53
Ethylene (C_2H_4)	368
Acetylene (C_2H_2)	3
Carbon Monoxide (CO)	7
Carbon Dioxide (CO_2)	361
Nitrogen (N_2)	86,027
Oxygen (O_2)	1,177
TDCG	500

Table 15 – Rogers Ratios for Key Gases

Code range of ratios		$\frac{C_2H_2}{C_2H_4}$	$\frac{CH_4}{H_2}$	$\frac{C_2H_4}{C_2H_6}$	Detection limits and 10 x detection limits are shown below:
					C_2H_2 1 ppm 10 ppm
					C_2H_4 1 ppm 10 ppm
					CH_4 1 ppm 10 ppm
					H_2 5 ppm 50 ppm
					C_2H_6 1 ppm 10 ppm
<0.1		0	1	0	
0.1-1		1	0	0	
1-3		1	2	1	
>3		2	2	2	
Case	**Fault Type**				**Problems Found**
0	No fault	0	0	0	Normal aging
1	Low energy partial discharge	1	1	0	Electric discharges in bubbles, caused by insulation voids or super gas saturation in oil or cavitation (from pumps) or high moisture in oil (water vapor bubbles).
2	High energy partial discharge	1	1	0	Same as above but leading to tracking or perforation of solid cellulose insulation by sparking, or arcing; this generally produces CO and CO_2.
3	Low energy discharges, sparking, arcing	1 - 2	0	1-2	Continuous sparking in oil between bad connections of different potential or to floating potential (poorly grounded shield, etc); breakdown of oil dielectric between solid insulation materials.
4	High energy discharges, arcing	1	0	2	Discharges (arcing) with power follow through; arcing breakdown of oil between windings or coils, between coils and ground, or load tap changer arcing across the contacts during switching with the oil leaking into the main tank.
5	Thermal fault less than 150 °C (see note 2)	0	0	1	Insulated conductor overheating; this generally produces CO and CO_2 because this type of fault generally involves cellulose insulation.
6	Thermal fault temperature range 150-300 °C (see note 3)	0	2	0	Spot overheating in the core due to flux concentrations. Items below are in order of increasing temperatures of hot spots: small hot spots in core; shorted laminations in core; overheating of copper conductor from eddy currents; bad connection on winding to incoming lead or bad contacts on load or no-load tap changer; circulating currents in core—this could be an extra core ground (circulating currents in the tank and core); this could also mean stray flux in the tank. These problems may involve cellulose insulation which will produce CO and CO_2.

Table 15 – Rogers Ratios for Key Gases (continued)

Case	Fault Type				Problems Found
7	Thermal fault temperature range 300-700 °C	0	2	1	
8	Thermal fault temperature range over 700 °C (see note 4)	0	2	2	

Notes:

1. There will be a tendency for ratio C_2H_2/C_2H_4 to rise from 0.1 to above 3 and the ratio C_2H_4/C_2H_6 to rise from 1-3 to above 3 as the spark increases in intensity. The code at the beginning stage will then be 1 0 1.

2. These gases come mainly from the decomposition of the cellulose which explains the zeros in this code.

3. This fault condition is normally indicated by increasing gas concentrations. CH_4/H_2 is normally about 1; the actual value above or below 1 is dependent on many factors, such as the oil preservation system (conservator, N_2 blanket, etc.), the oil temperature, and oil quality.

4. Increasing values of C_2H_2 (more than trace amounts) generally indicate a hot spot higher than 700 °C. This generally indicates arcing in the transformer. If acetylene is increasing, and especially if the generation rate is increasing, the transformer should be de-energized; further operation is extremely hazardous.

General Remarks:

1. Values quoted for ratios should be regarded as typical (not absolute). This means that the ratio numbers are not "carved in stone"; there may be transformers with the same problems whose ratio numbers fall outside the ratios shown at the top of the table.

2. Combinations of ratios not included in the above codes may occur in the field. If this occurs, the Rogers Ratio Method will not work for analyzing these cases.

3. Transformers with on-load tap changers may indicate faults of code type 2 0 2 or 1 0 2 depending on the amount of oil interchange between the tap changer tank and the main tank.

circulating currents in the tank and core). See the two bottom problems on table 16 (later in this chapter). This example was chosen to show a transformer that was not a "clear cut" diagnosis. Engineering judgment is always required.

A small quantity of acetylene is present—just above the detection limit of 1 ppm. This is not high energy arcing due to the small amount;

more likely, it has been produced by a one-time, nearby lightning strike or a voltage surge.

Rogers Ratio					
					Code
C_2H_2/C_2H_4	=	3/368	=	0.00815	0
CH_4/H_2	=	60/9	=	6.7	2
C_2H_4/C_2H_6	=	368/53	=	6.9	2

This code combination is Case 8 in table 15 which indicates that this transformer has a thermal fault hotter than 700 °C.

Example 2

	Latest DGA (ppm)	Prior DGA No. 2 (ppm)	Prior DGA No. 1 (ppm)
Hydrogen (H_2)	26	27	17
Methane (CH_4)	170	164	157
Ethane (C_2H_6)	278	278	156
Ethylene (C_2H_4)	25	4	17
Acetylene (C_2H_2)	2	0	0
Carbon Monoxide (CO)	92	90	96
Carbon Dioxide (CO_2)	3,125	2,331	2,476
Nitrogen (N_2)	67,175	72,237	62,641
Oxygen (O_2)	608	1,984	440

Rogers Ratio Analysis Based on Latest DGA					
					Codes
C_2H_2/C_2H_4	=	2/25	=	0.080	0
CH_4/H_2	=	170/2	=	6.54	2
C_2H_4/C_2H_6	=	25/27	=	0.09	0

Notice that methane is increasing slowly, but ethane had a large increase between samples 1 and 2 but did not increase between samples 2 and 3. Note that two key gases (CH_2 and C_2H_6) are above IEEE Condition 1 in table 9, so the Rogers Ratio Method is valid. By referring to table 15, this combination of codes is Case 6, which indicates the transformer has a thermal fault in the temperature range of 150 °C to 300 °C.

Life history of the transformer must be examined carefully. It is, again, very important to keep accurate records of every transformer. This information is invaluable when it becomes necessary to do an evaluation.

The transformer in this example is one of three sister transformers that have had increased cooling installed and are running higher loads due to a generator upgrade several years ago. Transformer sound level (hum) is markedly higher than for the two sister transformers. The unit breaker experienced a fault some years ago, which placed high mechanical stresses on the transformer. This generally means loose windings, which can generate gas due to friction (called a thermal fault) by Rogers Ratios. Comparison with sister units reveals almost triple the ethane as the other two, and it is above the IEEE Condition 4. Gases are increasing slowly; there has been no sudden rate increase in combustible gas production. Notice the large increase in O_2 and N_2 between the first and second DGA and the large decrease between the second and third. This indicates that the oil sample was exposed to air (atmosphere) and that these two gases are inaccurate in the middle sample.

6.1.10 Carbon Dioxide/Carbon Monoxide Ratio

This ratio is not included in the Rogers Ratio Method of analysis. However, it is useful to determine if a fault is affecting the cellulose insulation. This ratio is included in transformer oil analyzing software programs, such as Delta X Research Transformer Oil Analyst. This analysis is available from the TSC at D-8440 and D-8450 in Denver.

Formation of CO_2 and CO from the degradation of oil impregnated paper increases rapidly with temperature. Calculate a normal operating CO_2/CO ratio at each DGA, based on the total accumulated amount of both gases. Look at several DGAs concentrating on CO_2 and CO. Experience has shown that, with normal loading and temperatures, the CO_2 generation rates runs 7 to 20 times higher than CO. With a CO_2/CO ratio above 7, there is little concern. With some transformers, ratios down to 5 times more CO_2 than CO might be considered normal. However, be careful with a ratio below 7. If H_2, CH_4, and C_2H_6 are increasing significantly as well as CO and the ratio is 5 or less, there is probably a problem. Take time to know the particular transformer by carefully checking all prior DGAs and establishing a normal operating CO_2 to CO ratio.

CAUTION:

After a suspected problem (a substantial increase in the amount of CO), the ratio should be based on the gas generation of both CO_2 and CO between successive DGAs and not on accumulated total CO_2 and CO gas levels.

If a problem is suspected, immediately take another DGA sample to confirm the problem. Take the amount of CO_2 generated between the DGAs and divide it by the amount of CO generated at that same time to establish the ratio. An excellent indication of abnormally high temperatures and rapidly deteriorating cellulose insulation is a CO_2/CO under 5. If the ratio is 3 or under, severe and rapid deterioration of cellulose is certainly occurring. In addition to DGAs, perform the Furans test explained in the following section 7.6. Extreme overheating from loss of cooling or plugged oil passages will produce a CO_2/CO ratio around 2 or 3 along with increasing Furans. If this is found, de-energization and internal inspection is recommended; the transformer is in imminent danger of failure.

Table 16 is adapted from IEC 60599, Appendix A.1.1 [13]. Some of the wording has been changed to reflect American language usage rather than European.

Table 16 – Typical Faults in Power Transformers [13]

Fault	Examples
Partial discharges	Discharges in gas-filled cavities in insulation resulting from incomplete impregnation, high moisture in paper, gas in oil supersaturation or cavitation, (gas bubbles in oil) leading to X wax formation on paper.
Discharges of low energy	Sparking or arcing between bad connections of different floating potential, from shielding rings, toroids, adjacent discs or conductors of different windings, broken brazing, closed loops in the core. Additional core grounds. Discharges between clamping parts, bushing and tank, high voltage and ground, within windings. Tracking in wood blocks, glue of insulating beam, winding spacers. Dielectric breakdown of oil, load tap changer breaking contact.
Discharges of high energy	Flashover, tracking or arcing of high local energy or with power follow-through. Short circuits between low voltage and ground, connectors, windings, bushings, and tank, windings and core, copper bus and tank, in oil duct. Closed loops between two adjacent conductors around the main magnetic flux, insulated bolts of core, metal rings holding core legs.
Overheating less than 300 °C	Overloading the transformer in emergency situations. Blocked or restricted oil flow in windings. Other cooling problem, pumps valves, etc. See the "Cooling" section in this document. Stray flux in damping beams of yoke.
Overheating 300 to 700 °C	Defective contacts at bolted connections (especially busbar), contacts within tap changer, connections between cable and draw-rod of bushings. Circulating currents between yoke clamps and bolts, clamps and laminations, in ground wiring, bad welds or clamps in magnetic shields. Abraded insulation between adjacent parallel conductors in windings.
Overheating over 700 °C	Large circulating currents in tank and core. Minor currents in tank walls created by high uncompensated magnetic field. Shorted core laminations.

Notes:

1. X wax formation comes from Paraffinic oils (paraffin based). These are not used in transformers at present in the United States but are predominate in Europe.

2. The last overheating problem in the table says "over 700 °C." Recent laboratory discoveries have found that acetylene can be produced in trace amounts at 500 °C, which is not reflected in this table. We have several transformers that show trace amounts of acetylene that probably are not active arcing but are the result of high-temperature thermal faults, as in the example. It may also be the result of one arc, due to a nearby lightning strike or voltage surge.

3. A bad connection at the bottom of a bushing can be confirmed by comparing infrared scans of the top of the bushing with a sister bushing. When loaded, heat from a poor connection at the bottom will migrate to the top of the bushing, which will display a markedly higher temperature. If the top connection is checked and found tight, the problem is probably a bad connection at the bottom of the bushing.

6.1.11 Moisture Problems

Moisture, especially in the presence of oxygen, is extremely hazardous to transformer insulation. Recent EPRI studies show that an oxygen level above 2,000 ppm dissolved in transformer oil is extremely destructive. Each DGA and Doble test result should be examined carefully to see if water content is increasing and to determine the moisture by dry weight (M/DW) or percent saturation in the paper insulation. When 2% M/DW is reached, make plans for a dry out. Never allow the M/DW to go above 2.5% in the paper or 30% oil saturation without drying out the transformer. Each time the moisture is doubled in a transformer, the life of the insulation is cut by one-half. Keep in mind that the life of the transformer is the life of the paper, and the purpose of the paper is to keep out moisture and oxygen. For service-aged transformers rated less than 69 kV, results of up to 35 ppm at 60 °C are considered acceptable. For 69 kV through 230 kV, a DGA test result of 20 ppm at 60 °C is considered acceptable. For greater than 230 kV, moisture should never exceed 12 ppm at 60 °C. However, the use of absolute values for water does not always guarantee safe conditions, and the percent by dry weight should be determined. See table 19, "Doble Limits for In-Service Oils," in section 7.6. If values are higher, the oil should be processed. If the transformer is kept as dry and free of oxygen as possible, transformer life will be extended.

Reclamation specifies that manufacturers dry new transformers to no more than 0.5% M/DW during commissioning. That means that a transformer with 10,000 pounds of paper insulation, has 10,000 x 0.005 = 50 pounds of water (about 6 gallons) in the paper. This is not enough moisture to be detrimental to electrical integrity. When the transformer is new, this water is distributed equally through the transformer. It is extremely important to remove as much water as possible.

When the transformer is energized, water begins to migrate to the coolest part of the transformer and the site of the greatest electrical stress. This location is normally the insulation in the lower one-third of the winding [6]. Paper insulation has a much greater affinity for water than does the oil. The water will distribute itself unequally, with much more water being in the paper than in the oil. The paper will partially dry the oil by absorbing water out of the oil. Temperature is also a big factor in how the water distributes itself between the oil and paper. See table 17 below for comparison.

Table 17 – Comparison of Water Distribution in Oil and Paper [6]

Temperature (degrees C)	Water in Oil	Water in Paper
20	1	3,000 times what is in the oil
40	1	1,000 times what is in the oil
60	1	300 times what is in the oil

Table 17 shows the tremendous attraction that paper insulation has for water and how the water changes in the paper with temperature. The ppm of water in oil shown in the DGA is only a small part of the water in the transformer. **When an oil sample is taken, it is important to record the oil temperature from the top oil temperature gauge.**

Some laboratories give percent M/DW of the insulation in the DGA; others give percent oil saturation, and some give only the ppm of water in the oil. If you have an accurate temperature of the oil and the ppm of water, the Nomogram (figure 55, section 6.1.11.2) will give percent M/DW of the insulation and the percent oil saturation.

Where does the water come from? Moisture can be in the insulation when it is delivered from the factory. If the transformer is opened for inspection, the insulation can absorb moisture from the atmosphere. If there is a leak, moisture can enter in the form of water or humidity in the air. Moisture is also formed by the degradation of insulation as the transformer ages. Most water penetration is flow of wet air or rain water through poor gasket seals due to a pressure difference caused by transformer cooling. If a transformer is removed from service during rain or snow, some transformer designs cool rapidly and the pressure inside drops. The most common moisture ingress points are gaskets between bushing bottoms and the transformer top and the pressure relief device gasket. Small oil leaks, especially in the oil cooling piping, will also allow moisture ingress. With rapid cooling and the resultant pressure drop, relatively large amounts of water and water vapor can be pumped into the transformer in a short time. It is important to repair small oil leaks; the small amount of visible oil is not important in itself, but it indicates a point where moisture will enter [23].

It is critical for life extension to keep transformers as dry and as free of oxygen as possible. Moisture and oxygen cause the paper insulation to decay much faster than normal and form acids, sludge, and more moisture. Sludge settles on windings and inside the structure, causing less efficient transformer cooling, which allows temperature to slowly rise over a period of time. (This was discussed earlier in section section 3.4.5.4). Acids cause an increase in the rate of decay, which forms more acid, sludge, and moisture at a faster rate [21]. This is a vicious cycle of increasing speed forming more acid and causing more decay. The answer is to keep the transformer as dry as possible and as free of oxygen as possible. In addition, oxygen inhibitor should be watched in the DGA testing. The transformer oil should be dried when moisture reaches the values stated in table 19 (following later in this document). An inhibitor (ditertiary butyl paracresol [DBPC]) should be added (0.3% by weight ASTM D-3787) when the oil is processed (see section 7.3).

Water can exist in a transformer in five forms:

1. Free water, at the bottom of the tank.

2. Ice at the tank bottom (if the oil specific gravity is greater than 0.9, ice can float).

3. Water can be in the form of a water/oil emulsion.

4. Water can be dissolved in the oil and is given in ppm in the DGA.

5. Water can be in the form of humidity if transformers have an inert gas blanket.

Free water causes few problems with dielectric strength of oil; however, it should be drained as soon as possible. Having a water-oil interface allows oil to dissolve water and transport it to the insulation. Problems with moisture in insulation were discussed above. If the transformer is out of service in winter, water can freeze. If oil specific gravity is greater than 0.9 (ice specific gravity), ice will float. This can cause transformer failure if the transformer is energized with floating ice inside. This is one reason that DGA laboratories test the specific gravity of transformer oil.

The amount of moisture that can be dissolved in oil increases with temperature (see figure 51). This is why hot oil is used to dry out a transformer. A water/oil emulsion can be formed by purifying oil at too high temperature. When the oil cools, dissolved moisture forms an emulsion [21]. A water/oil emulsion causes drastic reduction in dielectric strength.

How much moisture in insulation is too much? When the insulation reaches 2.5% M/DW or 30% oil saturation (given on some DGAs), the transformer should have a dry out with vacuum, if the tank is rated for vacuum. If the transformer is old, pulling a vacuum can do more harm than good. In this case, it is better to do round-the-clock re-circulation with a Bowser drying the oil as much as possible, which will pull water out of the paper. At 2.5% M/DW, the paper insulation is degrading much faster than normal [6]. As the paper is degraded, more water is produced from the decay products, and the transformer becomes even wetter and decays even faster. When a transformer gets above 4% M/DW, it is in danger of flashover if the temperature rises to 90 °C.

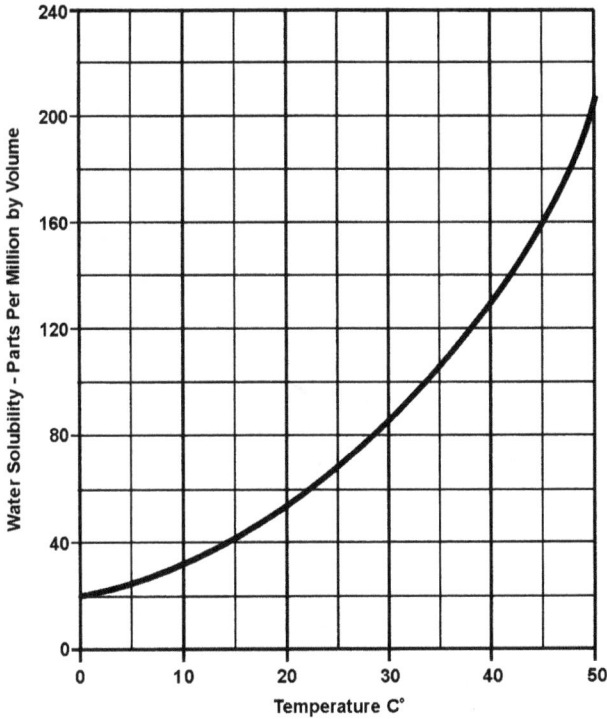

Figure 51 – Maximum Amount of Water Dissolved
in Mineral Oil Versus Temperature.

6.1.11.1 Dissolved Moisture in Transformer Oil – Moisture is
measured in the dissolved gas analysis in ppm. Some laboratories also
give percent saturation, which is the percent saturation of water in the
oil. This is a percentage of how much water is in the oil compared
with the maximum amount of water the oil can hold. Figure 51 shows
it can be seen that the amount of water the oil can dissolve is greatly
dependent on temperature. The curves below (figure 52) are percent
saturation curves. On the left line, find the ppm of water from your
DGA. From this point, draw a horizontal with a straight edge. From
the oil temperature, draw a vertical line. At the point where the lines
intersect, read the percent saturation curve. If the point falls between
two saturation curves, estimate the percent saturation based on where

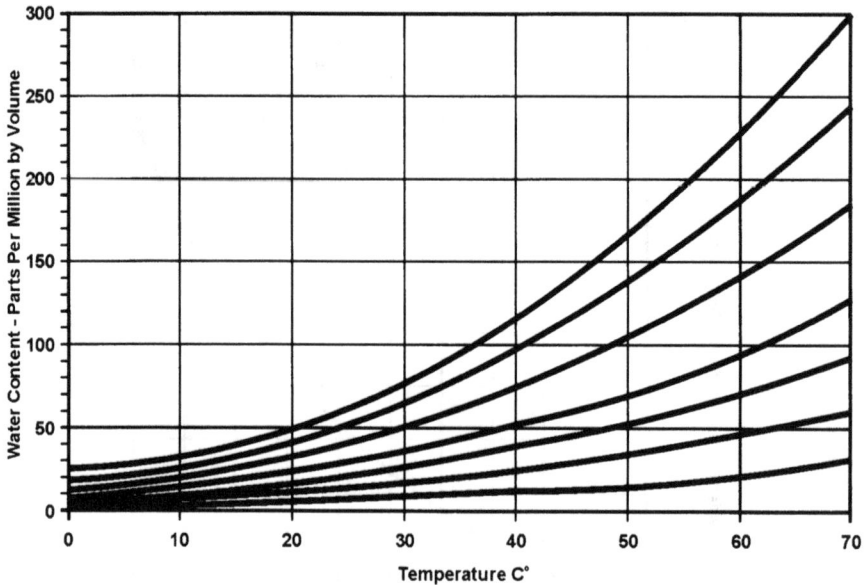

Figure 52 – Transformer Oil Percent Saturation Curves.

the point is located. For example, if the water is 30 ppm and the temperature is 40 °C, you can see on the curves that this point of intersection falls about halfway between the 20% curve and the 30% curve. This means that the oil is approximately 25% saturated. Curves shown on figure 52 are from IEEE 62-1995 [20].

CAUTION:

Below 30 °C, the curves are not very accurate.

6.1.11.2 Moisture in Transformer Insulation. Figure 53 shows how moisture is distributed throughout transformer insulation. Notice that the moisture is distributed according to temperature, with most moisture at the bottom and less moisture as temperature increases toward the top. This example shows almost twice the moisture near the bottom as there is at the top. Most service-aged transformers fail in the lower one-third of the windings, which is the area of most moisture. The area of most moisture is also the area of most electrical stress.

Moisture and oxygen are two of the transformer's worst enemies. It is very important to keep the insulation and oil as dry as possible and as free of oxygen as possible.

Failures due to moisture are the most common cause of transformer failures [6]. Without an accurate oil temperature, it is impossible for laboratories to provide accurate information about the M/DW or percent saturation. It will also be impossible for you to calculate this information accurately.

Figure 53 – Water Distribution in Transformer Insulation.

Experts disagree on how to tell how much moisture is in the insulation based on how much moisture is in the oil (ppm). At best, methods to determine moisture in the insulation based solely on DGA are inaccurate.

The methods discussed below to determine moisture levels in the insulation are estimates, and no decision should be made based on one DGA. However, keep in mind that the life of the transformer is the life of the insulation. The insulation is quickly degraded by excess moisture and the presence of oxygen. Base any decisions on several DGAs over a period of time and establish a trend of increasing moisture.

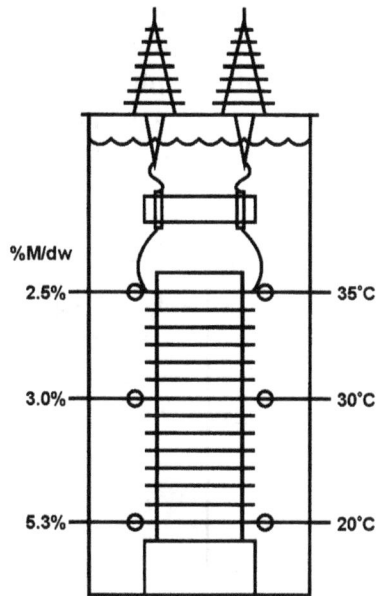

147

If the lab does not provide the percent M/DW, IEEE 62-1995 [20] gives a method. From the curve (figure 54), find the temperature of the **bottom oil sample and add 5 °C.** Do not use the top oil temperature. This estimates the temperature of the bottom third (coolest part) of the winding, where most of the water is located.

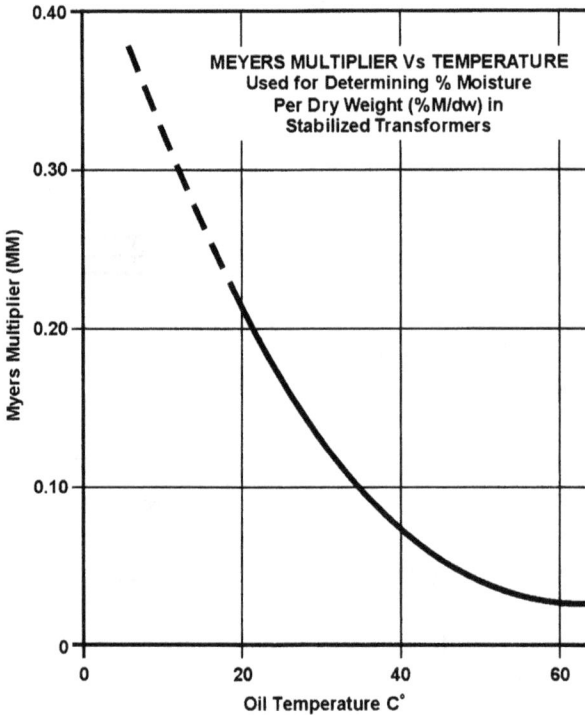

Figure 54 – Myers Multiplier Versus Temperature.

From this temperature, move up vertically to the curve. From this point on the curve, move horizontally to the left and find the Myers Multiplier number. Take this number and multiply the ppm of water shown on the DGA. The result is percent M/DW in the upper part of the insulation. This method gives less amount of water than the General Electric nomogram (figure 55).

This nomogram, published by General Electric in 1974, gives the percent saturation of oil and percent M/DW of insulation. Use the nomogram to check yourself after you have completed the method illustrated in figure 54. The nomograph in figure 55 will show more moisture than the IEEE method.

The curves in figure 55 are useful to help understand relationships between temperature, percent saturation of the oil, and percent M/DW of the insulation. For example, pick a point on the ppm water line (10 ppm). Place a straight edge on that point and pick a point on the temperature line (45 °C). Read the percent saturation

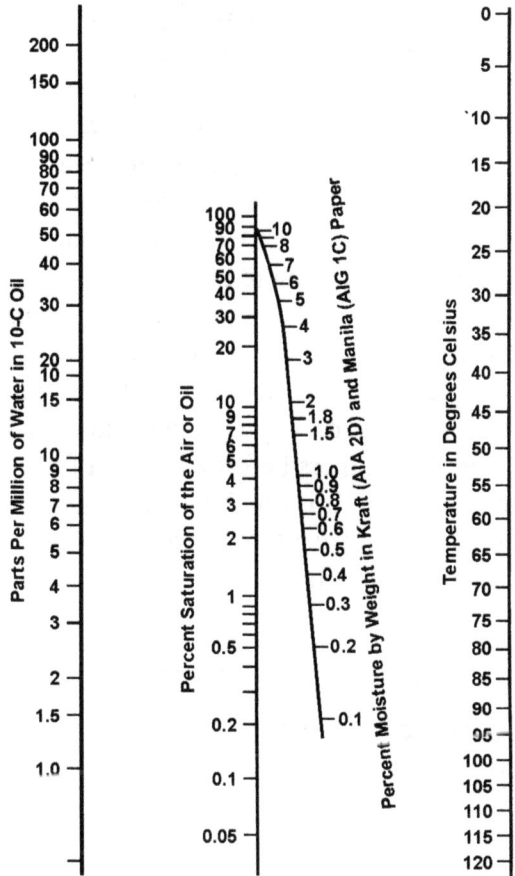

Nomogram for In-Service and Service-Aged Transformers for Determining Water Content of Paper Insulation and Oil.

Figure 55 – Water Content of Paper and Oil Nomogram.

and percent M/DW on the center lines. In this example, percent saturation is about 6.5%, and the % M/DW is about 1.5%. Now, hold the 10 ppm point and move the sample temperature upward (cooler) and notice how quickly the moisture numbers increase. For example, use 20 °C and read the percent saturation of oil at about 18.5% and the

% M/DW at about 3.75%. The cooler the oil, the higher the moisture percentage for the same ppm of water in the oil.

Do not make a decision on dry out based on only one DGA and one calculation; it should be based on trends over a period of time. Take additional samples and send them for analysis. Take extra care to make sure the oil temperature is correct. You can see by the nomogram that moisture content varies dramatically with temperature. Take extra care that the sample is not exposed to air. After using the more conservative IEEE method, if, again, subsequent samples show M/DW is 2.5% or more and the oil is 30% saturated or more, the transformer should be dried as soon as possible. Check the nomogram and curves above to determine the percent saturation of the oil. The insulation is degrading much faster than normal due to the high moisture content. Drying can be an expensive process; it is prudent to consult with others before making a final decision to implement dry out. However, it is much less expensive to perform a dry out than to allow a transformer to degrade faster than normal, substantially shortening transformer life.

7. Transformer Oil Tests That Should Be Completed Annually with the Dissolved Gas Analysis

7.1 Dielectric Strength
This test measures the voltage where the oil electrically breaks down. The test gives a good indication of the amount of contaminants (water and oxidation particles) in the oil. DGA laboratories typically use ASTM D-1816. **Using the D-1816 test, the minimum oil breakdown voltage is 20 kV for transformers rated less than 288 kV and 25 kV for transformers 287.5 kV and above.** If the dielectric strength test falls below these numbers, the oil should be reclaimed. Do not base any decision on one test result or on one type of test; instead, look at all the information over several DGAs and establish trends before making any decision. The dielectric strength

test is not extremely valuable; moisture in combination with oxygen and heat will destroy cellulose insulation long before the dielectric strength of the oil has given a clue that anything is going wrong [6]. The dielectric strength test also reveals nothing about acids and sludge. The tests explained below are much more important.

7.1.1 Interfacial Tension

This test (ASTM D-791-91 [22]), is used by DGA laboratories to determine the interfacial tension between the oil sample and distilled water. The oil sample is put into a beaker of distilled water at a temperature of 25 °C. The oil should float because its specific gravity is less than that of water (specific gravity of water is one). There should be a distinct line between the two liquids. The IFT number is the amount of force (dynes) required to pull a small wire ring upward a distance of 1 centimeter through the water/oil interface. (A dyne is a very small unit of force equal to 0.000002247 pound.) Good, clean oil will make a very distinct line on top of the water and give an IFT number of 40 to 50 dynes per centimeter of travel of the wire.

As the oil ages, it is contaminated by tiny particles (oxidation products) of the oil and paper insulation. Particles on top of the water extend across the water/oil interface line that weaken the surface tension between the two liquids. Particles in oil weaken interfacial tension and lower the IFT number. The IFT and acid numbers, together, are an excellent indication of when the oil needs to be reclaimed. It is recommended that the oil be reclaimed when the IFT number falls to 25 dynes per centimeter. At this level, the oil is very contaminated and must be reclaimed to prevent sludging, which begins around 22 dynes per centimeter. See *FIST 3-5* [21].

If oil is not reclaimed, sludge will settle on windings, insulation, cooling surfaces, etc., and cause loading and cooling problems, as discussed earlier. This will greatly shorten transformer life.

A definite relationship exists between the acid number, the IFT, and the number of years in service. The accompanying curve (figure 56)

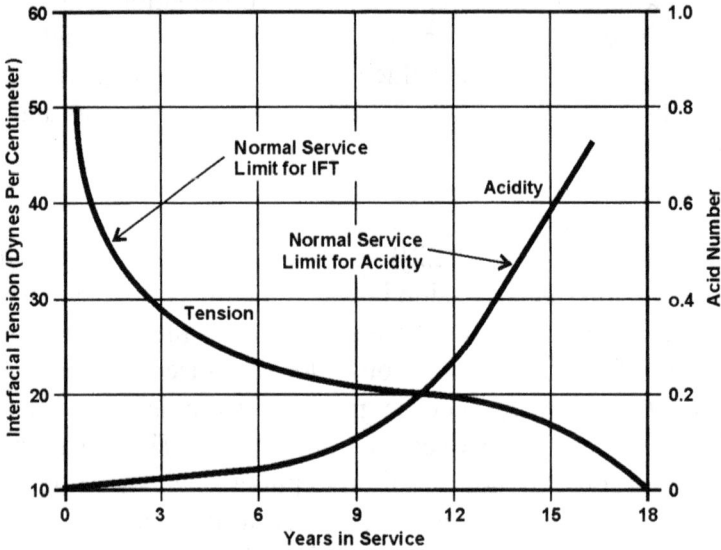

Figure 56 – Interfacial Tension, Acid Number,
Years in Service.

shows the relationship and is found in many publications. (It was originally published in the AIEE transactions in 1955.) Notice that the curve shows the normal service limits for both the IFT and the acid number.

7.2 Acid Number

Acid number (acidity) is the amount of potassium hydroxide (KOH) in milligrams (mg) that it takes to neutralize the acid in 1 gram (gm) of transformer oil. The higher the acid number, the more acid is in the oil. New transformer oils contain practically no acid. Oxidation of the insulation and oils forms acids as the transformer ages. The oxidation products form sludge and precipitate out inside the transformer. The acids attack metals inside the tank and form soaps (more sludge). Acid also attacks cellulose and accelerates insulation degradation. Sludging begins when the acid number reaches 0.40; it is obvious that the oil should be reclaimed before it reaches 0.40. It is recommended that the oil be reclaimed when it reaches 0.20 mg potassium hydroxide per gram (KOH/gm) [21]. As with all others, this decision must not be based on one DGA test, but watch for a rising trend in the acid number each year. Plan ahead and begin budget planning before the acid number reaches 0.20.

7.3 Test for Oxygen Inhibitor

In previous sections, the need to keep the transformer dry and O_2 free was emphasized. Moisture is destructive to cellulose, and even more so in the presence of oxygen. Some publications state that each time you double the moisture (ppm), you halve the life of the transformer. As was discussed, acids are formed that attack the insulation and metals that form more acids, causing a viscous cycle. Oxygen inhibitor is a key to extending the life of transformers. The inhibitor currently used is ditertiary butyl paracresol (DBPC). This works similar to a sacrificial anode in grounding circuits. The oxygen attacks the inhibitor instead of the cellulose insulation. As this occurs and the transformer ages, the inhibitor is used up and needs to be replaced. Replacement of the inhibitor also generally requires treatment of the oil [6]. The ideal amount of DBPC is 0.3% by total weight of the oil (shown on the transformer nameplate) [30].

Test the inhibitor content with the DGA every 3 to 5 years. If the inhibitor is 0.08%, the transformer is considered uninhibited, and the oxygen freely attacks the cellulose. If the inhibitor falls to 0.1%, the transformer should be re-inhibited. For example, if your transformer tested 0.1%, you need to go to 0.3% by adding 0.2% of the total weight of the transformer oil. The nameplate gives the weight of oil— for example, 5,000 pounds—so 5,000 pounds x 0.002 = 10 pounds of DBPC needs to be added. Adding a little too much DBPC does not hurt the oil. Dissolve 10 pounds of DBPC in the transformer oil that you have heated to the same temperature as the oil inside the transformer. It may take experimenting some to get the right amount of oil to dissolve the DBPC. Mix the oil and inhibitor in a clean container until all the DBPC is dissolved. Add this mixture to the transformer, using the method given in the transformer's instruction manual for adding oil.

CAUTION:

Do not attempt this unless you have had experience. Contact an experienced contractor or experienced Reclamation personnel if you need help.

In either case, do not neglect this important maintenance function; it is critical to transformer insulation to have the proper amount of oxygen inhibitor.

7.4 Power Factor

Power factor indicates the dielectric loss (leakage current) of the oil. This test may be completed by the DGA laboratories. It may also be completed by Doble testing. A high power factor indicates deterioration and/or contamination by products, such as water, carbon, or other conducting particles; metal soaps caused by acids (formed as mentioned above); attacking transformer metals; and products of oxidation. The DGA labs normally test the power factor at 25 °C and 100 °C. Doble information [25] indicates that the inservice limit for power factor is less than 0.5% at 25 °C. If the power factor is greater than 0.5% and less than 1.0%, further investigation is required; the oil may require replacement or fullers earth filtering. If the power factor is greater than 1.0% at 25 °C, the oil may cause failure of the transformer; replacement or reclaiming is required. Above 2%, oil should be removed from service and replaced because equipment failure is imminent. The oil cannot be reclaimed.

7.5 Oxygen

Oxygen must be watched closely in DGA tests. Many experts and organizations, including EPRI, believe that oxygen levels in the oil above 2,000 ppm greatly accelerate paper deterioration. This becomes even more critical with moisture above safe levels. Under the same temperature conditions, cellulose insulation in low oxygen oil will last 10 times longer than insulation in high oxygen oil [6]. It is recommended that if oxygen reaches 10,000 ppm in the DGA, the oil should be de-gassed and new oxygen inhibitor installed. High atmospheric gases (O_2 and N_2) normally mean that a leak has developed in a bladder or diaphragm in the conservator. If there is no conservator and pressurized nitrogen is on top of the oil, expect to see high nitrogen but not high oxygen. Oxygen comes only from leaks and from deteriorating insulation.

7.6 Furans

When cellulose insulation decomposes due to overheating, organic compounds, in addition to CO_2 and CO, are released and dissolved in the oil. These chemical compounds are known as furanic compounds or furans. The most important one, for our purposes, is 2-furfuraldehyde. When DGAs are required, always request that furans testing be completed by the laboratory to check for paper deterioration. In healthy transformers, there are no detectable furans in the oil, or they are less than 100 parts per billion (ppb). In cases where significant damage to paper insulation from heat has occurred, furan levels may be at least 100 ppb and up to 70,000 ppb. Furanic content in the oil is especially helpful in estimating remaining life in the paper insulation, particularly if several prior tests can be compared and trends established.

Use the furan numbers in table 18 for assessment; do not base any evaluation on only one test; use several DGAs over a period of time to develop trending. See *An Introduction to the Half-Century Transformer* by the Transformer Maintenance Institute, S.D. Myers Co., 2002 [39]. The first column in table 18 is used for transformers

Table 18 – Furans, DP, Percent of Life Used of Paper Insulation

55 °C Rise Transformer 2FAL (ppb)	65 °C Rise Trans-former Total Furans (ppb)	Estimated Degree of Polymerization (DP)	Estimated Percentage of Remaining Life	Interpretation
58	51	800	100	Normal Aging Rate
130	100	700	90	
292	195	600	79	
654	381	500	66	Accelerated Aging Rate
1,464	745	400	50	
1,720	852	380	46	
2,021	974	360	42	
2,374	1,113	340	38	Excessive Aging Danger Zone
2,789	1,273	320	33	
3,277	1,455	300	29	
3,851	1,664	280	24	High Risk of Failure
4,524	1,902	260	19	
5,315	2,175	240	13	End of Expected Life of Paper Insulation and of the Transformer
6,245	2,487	220	7	
7,337	2,843	200	0	

with non-thermally upgraded paper, and the second column is for transformers with thermally upgraded paper.

Testing is completed for five different furans which are caused by different problems. The five furans and their most common causes are listed below:

♦ 5H2F (5-hydroxymethyl-2-furaldehyde) caused by oxidation (aging and heating) of the paper

♦ 2FOL (2-furfurol) caused by high moisture in the paper

♦ 2FAL (2-furaldehyde) caused by overheating

♦ 2ACF (2-acetylfuran) caused by lightning (rarely found in DGA)

♦ 5M2F (5-methyl-2-furaldehyde) caused by local severe overheating (hotspot)

Doble inservice limits are reproduced below to support the above recommended guidelines.

Table 19 is excerpted from Doble Engineering Company's *Reference Book on Insulating Liquids and Gases* [25]. These Doble Oil Limit tables support information given in prior sections and are shown here as summary tables.

Additional guidelines given in table 20 are useful.

Table 19 – Doble Limits for Inservice Oils

	Voltage Class		
	69 kV	>69 288 kV	>288 kV
Dielectric Breakdown Voltage, D 877, kV minimum	26	30	[1]
Dielectric Breakdown Voltage D 1816, .04-inch gap, kV, minimum	20	20	25
Power Factor at 25 °C, D 924, maximum	0.5	0.5	0.5
Water Content, D 1533, ppm, maximum	[2]35	[2]25	[2]20
Interfacial Tension, D 971, dynes per centimeter (dynes/cm), minimum	25	25	25
Neutralization Number, D 974, mg KOH/gm, maximum	0.2	0.15	0.15
Visual Exam	clear and bright	clear and bright	clear
Soluble Sludge	ND[3]	ND[3]	ND[3]

[1] D 877 test is not as sensitive to dissolved water as the D 1816 test and should not be used with oils for extra high-voltage (EHV) equipment. Dielectric breakdown tests do not replace specific tests for water content..

[2] The use of absolute values of water-in-oil (ppm) do not always guarantee safe conditions in electrical apparatus. The percent by dry weight should be determined from the curves provided. See the information in section 6.1.11.2.

[3] ND = None detectable.

These recommended limits for inservice oils are not intended to be used as absolute requirements for removing oil from service but to provide guidelines to aid in determining when remedial action is most beneficial. Remedial action will vary depending upon the test results. Reconditioning of oil, that is, particulate removal (filtration) and drying, may be required if the dielectric breakdown voltage or water content do not meet these limits. Reclamation (clay filtration) or replacement of the oil may be required if test values for power factor, interfacial tension, neutralization number, or soluble sludge do not meet recommended limits.

Table 20 – Additional Guidelines for Inservice Oils

Power factor at 25 °C		
< 0.5%		Acceptable
> 0.5 < 1.0%		Investigate, oil may require replacement or clay treatment
> 1.0 < 2.0%		Investigate, oil may cause failure of the equipment, oil may require replacement or clay treatment
> 2.0%		Remove from service, investigate, oil may require replacement or clay treatment
Neut. No. (mg KOH/gm)	IFT (dynes/cm)	
< 0.05	25	Acceptable
0.05 < 0.15	22 < 25	Clay treat or replace at convenience, 345 kV, clay treat or replace in immediate future
0.15 < 0.5	16 > 22	Clay treat or replace in immediate future
0.5	<16	Replace[1]

[1] When an oil is allowed to sludge in service, special treatment may be required to clean the core, coil, and tank.

7.7 Oil Treatment Specifications

After the oil is treated, the results should be as follows.

Gases		Physical Properties	
H_2	5 ppm or less	Water	less than 10 ppm
CH_4	5 ppm or less	IFT	40 dynes/cm minimum
C_2H_2	0 ppm	Acid number	0.03 mg KOH/gm maximum
CO	20 ppm or less	O_2 inhibitor	0.3% by oil weight
CO_2	300 ppm or less		
O_2	4,000 ppm or less		

Dielectric strength, ASTM D1816-97, kV minimum,	≥ 69 kV	> 69 to < 230 kV	230 kV and above
1-mm[1] gap (0.04-in)	23	28	30
2-mm gap (0.08-in)	40	47	50

[1] millimeter (mm).

7.7.1 Taking Oil Samples for DGA

Sampling procedures and lab handling are usually areas that cause the most problems in getting an accurate DGA. There are times when atmospheric gases, moisture, or hydrogen take a sudden leap from one DGA to the next. As has been mentioned, at these times, one should immediately take another sample to confirm DGA values. It is, of course, possible that the transformer has developed an atmospheric leak or that a fault has suddenly occurred inside. More often, the sample has not been taken properly, or it has been contaminated with atmospheric gases or mishandled in other ways. The sample must be protected from all contamination, including atmospheric exposure.

Do not take samples from the small sample ports located on the side of the large sample (drain) valves. These ports are too small to adequately flush the large valve and pipe nipple connected to the tank; in addition, air can be drawn past the threads and contaminate the sample. Fluid in the valve and pipe nipple remain dormant during operation and can be contaminated with moisture, microscopic stem packing particles, and other particles. The volume of oil in this location can also be contaminated with gases, especially hydrogen. Hydrogen is one of the easiest gases to form. With hot sun on the side of the transformer tank where the sample valve is located, high ambient temperature, high oil temperature, and captured oil in the sample valve and extension, hydrogen formed will stay in this area until a sample is drawn.

The large sample (drain) valve can also be contaminated with hydrogen by galvanic action of dissimilar metals. Sample valves are usually brass; a brass pipe plug should be installed when the valve is not being used. If a galvanized or black iron pipe plug is installed in a brass valve, the dissimilar metals produce a thermocouple effect, and circulating currents are produced. As a result, hydrogen is generated in the void between the plug and valve gate. If the valve is not thoroughly flushed, the DGA will show a high hydrogen level.

Oil should not be sampled for DGA purposes when the transformer is at or below a freezing temperature. Test values which are affected by water (such as dielectric strength, power factor, and dissolved moisture content) will be inaccurate.

CAUTION:

Transformers must not be sampled if there is a negative pressure (vacuum) at the sample valve.

This is typically not a problem with conservator transformers. If the transformer is nitrogen blanketed, look at the pressure/vacuum gauge. If the pressure is positive, go ahead and take the sample. If the pressure is negative, a vacuum exists at the top of the transformer. If there is a vacuum at the bottom, air will be pulled in when the sample valve is opened. Wait until the pressure gauge reads positive before sampling. **Pulling in a volume of air could be disastrous if the transformer is energized.**

If negative pressure (vacuum) is not too high, the weight of oil (head) will make positive pressure at the sample valve, and it will be safe to take a sample. Oil head is about 2.9 feet (2 feet 10.8 inches) of oil per psi. If it is important to take the sample even with a vacuum showing at the top, proceed as described below.

Use the sample tubing and adaptors described below to adapt the large sample valve to ⅛-inch tygon tubing. Fill a length (2 to 3 feet) of tygon tubing with new transformer oil (no air bubbles) and attach one end to the pipe plug and the other end to the small valve. Open the large sample (drain) valve a small amount and very slowly crack open the small valve. **If oil in the tygon tubing moves toward the transformer, shut off the valves immediately. Do not allow air to be pulled into the transformer.** If oil moves toward the transformer, there is a vacuum at the sample valve. Wait until the pressure is positive before taking the DGA sample. If oil is pushed out of the tygon tubing into the waste container, there is a positive pressure, and it is safe to proceed with DGA sampling. Shut off the valves and configure the tubing and valves to take the sample per the instructions below.

7.7.1.1 DGA Oil Sample Container – Glass sample syringes are recommended. There are different containers, such as stainless steel vacuum bottles and others. Using only glass syringes is recommended. If there is a small leak in the sampling tubing or connections, vacuum bottles will draw air into the sample, which cannot be seen inside the bottle. The sample will show high atmospheric gases and high moisture if the air is humid. Other contaminates such as suspended solids or free water cannot be seen inside the vacuum bottle. Glass syringes are the simplest to use because air bubbles are easily seen and expelled. Other contaminates are easily seen, and another sample can be immediately taken if the sample is contaminated. The downside is that glass syringes must be handled carefully and must be protected from direct sunlight. They should be returned to their shipping container immediately after taking a sample. If they are exposed to sunlight for any time, hydrogen will be generated, and the DGA will show false hydrogen readings.

For these reasons, glass syringes are recommended, and the instructions below include only this sampling method.

Obtain a brass pipe plug (normally 2 inches) that will thread into the sample valve at the bottom of the transformer. Drill and tap the pipe plug for ⅛-inch national pipe thread, insert a ⅛-inch pipe nipple (brass if possible) and attach a small ⅛-inch valve for controlling the sample flow. Attach a ⅛-inch tygon tubing adaptor to the small valve outlet. Sizes of the piping and threads above do not matter; any arrangement with a small sample valve and adaptor to ⅛-inch tygon tubing will suffice. See figure 57.

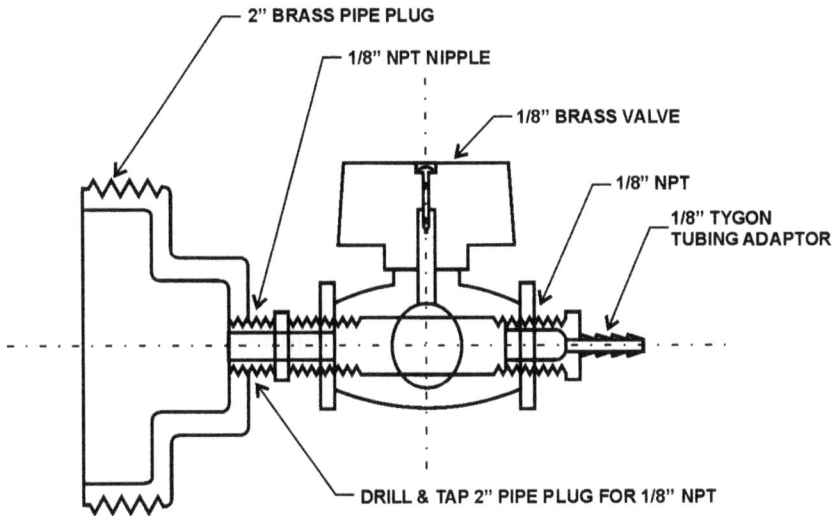

Figure 57 – Oil Sampling Piping.

7.7.1.2 *Taking the Sample*

♦ Remove the existing pipe plug and inspect the valve opening for rust and debris.

♦ Crack open the valve and allow just enough oil to flow into the waste container to flush the valve and threads. Close the valve and wipe the threads and outlet with a clean dry cloth.

♦ Re-open the valve slightly and flush approximately 1 quart of oil into the waste container.

♦ Install the brass pipe plug (described above) and associated ⅛-inch pipe and small valve, and a short piece of new ⅛-inch tygon tubing to the outlet of the ⅛-inch valve.

♦ Never use the same sample tubing on different transformers. This is one way a sample can be contaminated and give false readings.

♦ Open both the large valve and small sample valve and allow another quart to flush through the sampling apparatus. Close both valves. Do this before attaching the glass sample syringe. Make sure the short piece of tygon tubing that will attach to the sample syringe is installed on the ⅛-inch valve before you do this.

♦ Install the glass sample syringe on the short piece of ⅛-inch tubing. Turn the stopcock handle on the syringe so that the handle points toward the syringe. **Note: The handle always points toward the closed port.** The other two ports are open to each other. See figure 58.

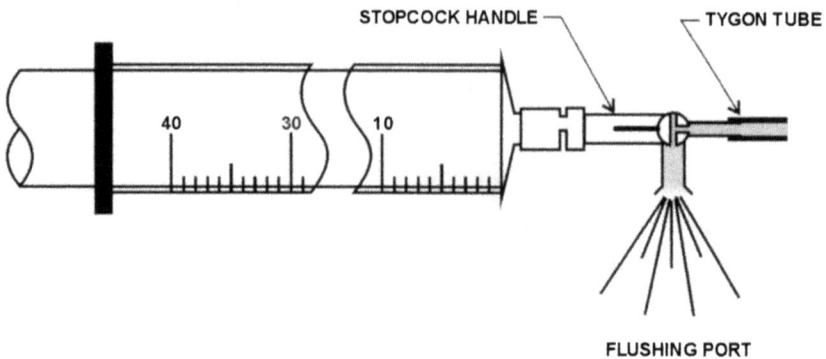

Figure 58 – Sampling Syringe (Flushing).

♦ Open the large sample valve a small amount and adjust the ⅛-inch valve so that a gentle flow goes through the flushing port of the glass syringe into the waste bucket.

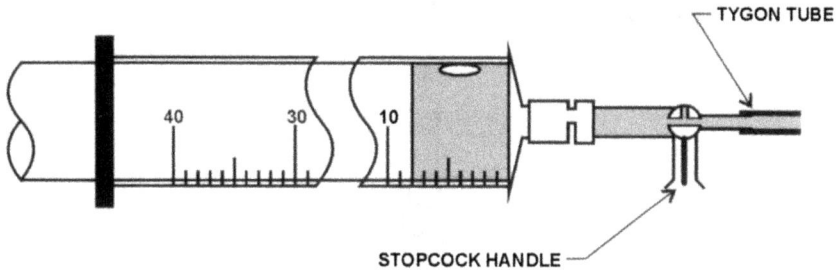

Figure 59 – Sampling Syringe (Filling).

♦ Slowly turn the syringe stopcock handle so that the handle points to the flushing port (figure 59). This closes the flushing and allows oil to flow into the sample syringe. Do not pull the syringe handle; this will create a vacuum and allow bubbles to form. The syringe handle (piston) should back out very slowly. If it moves too fast, adjust the small ⅛-inch valve until the syringe slows and hold your hand on the back of the piston so you can control the travel.

♦ Allow a small amount, about 10 cubic centimeters (cc), to flow into the syringe and turn the stopcock handle again so that it points to the syringe. This will again allow oil to come out of the flushing port into the waste bucket.

♦ Pull the syringe off the tubing, but do not shut off the oil flow. Allow the oil flow to continue into the waste bucket.

♦ Hold the syringe vertical and turn the stopcock up so that the handle points away from the syringe. Press the syringe piston to eject any air bubbles, but leave 1 or 2 cc of oil in the syringe. See figure 60.

CAUTION:

Do not eject all the oil, or air will re-enter.

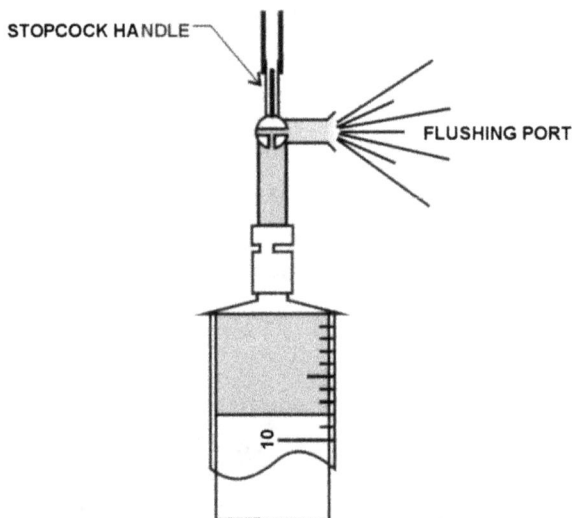

Figure 60 – Sample Syringe Bubble Removal.

♦ Turn the stopcock handle toward the syringe. The small amount of oil in the syringe should be free of bubbles and ready to receive the sample. If there are still bubbles at the top, repeat the process until you have a small amount of oil in the syringe with no bubbles.

♦ Reattach the tygon tubing. This will again allow oil to flow out of the flushing port. Slowly turn the stopcock handle toward the flushing port which again will allow oil to fill the syringe. The syringe piston will again back slowly out of the syringe. Allow the syringe to fill about 80% full. Hold the piston so you can stop its movement at about 80% full.

CAUTION:

Do not pull the piston. This will cause bubbles to form.

♦ Close the stopcock by turning the handle toward the syringe. Oil again will flow into the waste container. Shut off both valves, remove the sampling apparatus, and reinstall the original pipe plug.

CAUTION:

Do not eject any bubbles that form after the sample is collected; these are gases that should be included in the lab sample.

♦ Return the syringe to its original container immediately. Do not allow sunlight to impact the container for any length of time. Hydrogen will form and give false readings in the DGA.

♦ Carefully package the syringe in the same manner that it was shipped to the facility and send it to the lab for processing.

♦ Dispose of waste oil in the plant waste oil container.

8. Silicone Oil-Filled Transformers

8.1 Background

Silicone oils became more common when polychlorinated biphenols were discontinued. They are mainly used in transformers inside buildings and are smaller than generator step-up transformers. Silicone oils have a higher fire point than mineral oils and, therefore, are used where fire concerns are more critical. As of this writing, there are no definitive published standards. IEEE has a guide, and Doble has some service limits, but there are no standards. Information below is taken from the IEEE publication, from Doble, from articles, from IEC 60599 concepts, and from Delta X Research's/Transformer Oil Analyst rules. Silicone oil dissolved gas analysis is in the beginning stage, and the suggested methods and limits below are subject to change as we gain more experience. However, in the absence of any other methods and limits, use the ones below as a beginning.

Silicone oils used in transformers are polydimethylsiloxane fluids, which are different than mineral oils. Many of the gases generated by thermal and electrical faults are the same. The gases are generated in different proportions than with transformer mineral oils. Also, some fault gases have different solubilities in silicone oils than in mineral oils. Therefore, the same faults would produce different concentrations and different generation rates in silicone oils than mineral oils.

As with mineral oil-filled transformers, three principal causes of gas generation are aging, thermal faults, and/or electrical faults resulting in deterioration of solid insulation and deterioration of silicone fluid.

Overheating of silicone oils causes degradation of fluid and generation of gases. Generated gases depend on the amount of dissolved oxygen in the fluid, temperature, and how close bare copper conductors are to the heating. When a transformer is new, silicone oil will typically contain a lot of oxygen. Silicone transformers are typically sealed and pressurized with nitrogen. New silicone oil is not de-gassed; and, as a rule, oxygen concentration will be equivalent to oxygen solubility (maximum) in silicone. The silicone has been exposed to atmosphere for some time during manufacture of the transformer and manufacturer and storage of silicone oil itself. Therefore, carbon monoxide and carbon dioxide are easily formed and dissolved in the silicone due to the abundance of oxygen in the oil, resulting from this atmospheric exposure. In normal new silicone transformers (no faults), both carbon monoxide and carbon dioxide will be generated in the initial years of operation. As the transformer ages and oxygen is depleted, generation of these gases slows, and concentrations level off [27]. See figure 61 for the relationship of decreasing oxygen and increasing carbon monoxide and carbon dioxide as a transformer ages. This curve is for general information only and should not be taken to represent any

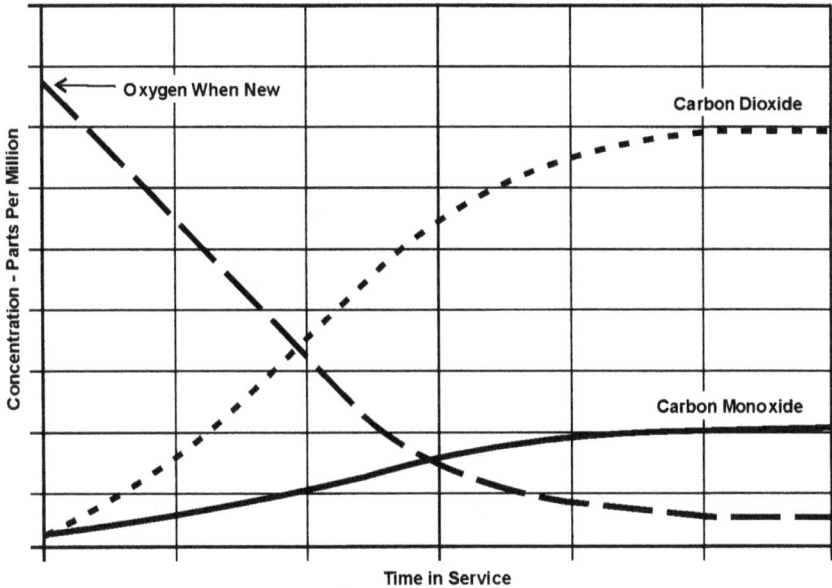

Figure 61 – Relationship of Oxygen to Carbon Dioxide and
Carbon Monoxide as Transformer Ages.

particular transformer. A real transformer with changes in loading,
ambient temperatures, and various duty cycles would make these
curves look totally different.

After the transformer is older (assuming no faults have occurred),
oxygen concentration will reach equilibrium (figure 61). Reaching
equilibrium may take a few years, depending on the size of the
transformer, loading, ambient temperatures, etc. After this time,
oxygen, carbon monoxide, and carbon dioxide level off; and the
rate of production of these gases from normal aging should be
relatively constant. If generation rates of these gases change greatly
(seen from the DGA), a fault has occurred, either thermal or electrical.
Rate of generation of these gases and amounts can be used to roughly
determine what the fault is. Once you notice a significant increase
in rate of generation of any gas, it is a good idea to subtract the
amount of gas that was already in the transformer before this

increase. This ensures that gases used in the diagnosis are only gases that were generated after the fault began.

8.2 Carbon Monoxide in Silicone Transformers

Carbon monoxide will be a lot higher in a silicone transformer than a mineral oil-filled one. It is difficult to try to determine what is producing the CO. Is it coming from normal aging of oil or from deterioration of paper from a fault condition? The only solution is a furan analysis. If the CO content is greater than the IEEE limit of 3,000 ppm [28] (table 19, shown later in this manual), and the generation rate G1 is met or exceeded, a furan analysis is recommended with the annual DGA. If a thermal fault is occurring and is producing CO and small amounts of methane and hydrogen, the fault may be masked by the normal production of CO from the silicone oil itself. If the CO generation rate has greatly increased, along with other gases, it becomes obvious that a fault has occurred. The furan analysis can only tell if the paper is involved (being heated) in the fault.

8.3 Comparison of Silicone Oil and Mineral Oil Transformers

Some general conclusions can be drawn by comparing silicone oil and mineral oil transformers.

1. All silicone oil-filled transformers will have a great deal more CO than normal mineral oil-filled transformers. CO can come from two sources—the oil itself and from degradation of paper insulation. If the DGA shows little other fault in gas generation besides CO, the only way to tell for certain if CO is coming from paper degradation (a fault) is to run a furan analysis with the DGA. If other fault gases are also being generated in significant amounts, in addition to CO, there obviously is a fault, and CO is coming from paper degradation.

2.   There will generally be more hydrogen present than in a mineral oil-filled transformer.

3.   Due to "fault masking," mentioned above, it is almost impossible to diagnose what is going on inside a silicone-filled transformer based solely on DGA. One exception is that if acetylene is being generated, there is an active arc. Look also at gas generation rates and operating history. Look at loading history, through faults, and other incidents. It is imperative that detailed records of silicone oil-filled transformers be carefully kept up to date. These are invaluable when a problem is encountered.

4.   If acetylene is being generated, there is a definitely an active electrical arc. The transformer should be removed from service.

5.   In general, oxygen in a silicone-filled transformer comes from atmospheric leaks or was present in the transformer oil when it was new. This oxygen is consumed as CO and CO2 are formed from the normal heating from operation of the transformer.

6.   Once the transformer has matured and the oxygen has leveled off and remained relatively constant for two or more DGA samples, if you see a sudden increase in oxygen, and perhaps carbon dioxide and nitrogen, the transformer has developed a leak.

8.4 Gas Limits

Table 21 shows IEEE limits [28], compared with Doble [27] in a study of 299 operating transformers. The table of gases from the Doble study seems more realistic, showing gas level average of 95% of transformers in the study. Note, with the last four gases, limits given by the IEEE (trial use guide) run over 70% higher than the Doble 95% norms. But with the first three gases, hydrogen, methane, and ethane, the IEEE limits are well below the amount of gas found in 95% norms

Table 21 – Comparison of Gas Limits

Gas	Doble 95% Norm	IEEE Limits
Hydrogen	511	200
Methane	134	100
Ethane	26	30
Ethylene	17	30
Acetylene	0.6	1
CO	1,749	3,000
CO2	15,485	30,000
Total Combustibles	2,024	3,360

in the Doble study. We obviously cannot have limits that are below the amount of gas found in normal operating transformers. Therefore, it is suggested that we use the Doble (95% norm) limits. The 95% norm limit means that 95% of the silicone oil transformers studied had gas levels below these limits. Obviously, 5% had gases higher than these limits. These are problem transformers that we should pay more attention to.

In table 22, the IEEE limits for L1 were chosen. For L2 limits, a statistical analysis was applied, and two standard deviations were added to L1 to obtain L2. For L3 limits, the L1 limits were doubled.

Limits L1, L2, and L3 represent the concentration in individual gases in ppm. G1 and G2 represent generation rates of individual gases in ppm per month. To obtain G1 and G2 in ppm per day, divide the per month numbers by 30. Except for acetylene, G1 is 10% of L1, and G2 is 50% of L1. The generation rates (G1, G2) are points where our level of concern should increase, especially when considered with the L1, L2, and L3 limits. At G2 generation rate, we should be extremely concerned, reduce the DGA sampling interval accordingly, and perhaps plan an outage, etc.

Table 22 – Suggested Levels of Concern (Limits)

Gas	L1 (ppm)	L2 (ppm)	L3 (ppm)	G1 (ppm per month)	G2 (ppm per month)
Hydrogen	200	240	400	20	100
Methane	100	125	200	10	50
Ethane	30	40	60	3	15
Ethylene	30	25	60	3	15
Acetylene	1	2	3	1	1
CO	3,000	3,450	6,000	300	1,500
CO2	30,000	34,200	60,000	1,500	15,000
TDCG	3,360	3,882	6,723	NA	NA

Except for acetylene, generation rate levels G1 and G2 were taken from IEC 60599 [13] which is used with mineral oil transformers. **Any amount of ongoing acetylene generation means active arcing inside the transformer. In this case, the transformer should be removed from service.** These criteria were chosen because of an absence of any other criteria. As DGA criteria for silicone oils becomes better known and the quantified table 22 will change to reflect new information.

As with mineral oil-filled transformers, gas generation rates are much more important than the amount of gas present. Total accumulated gas depends greatly on age (an older transformer has more gas). If the rate of generation of any combustible gas shows a sudden increase in the DGA, immediately take another oil sample to confirm the gas generation rate increase. If the second DGA confirms a generation rate increase, get some outside advice. Be careful; gas generation rates increase somewhat with temperature variations caused by increased loading and summer ambient temperatures. However, higher operating temperatures are also the most likely conditions for a fault to occur. The real question is, has the increased gas generation rate been caused by a fault or increased temperature from greater loading or higher ambient temperature?

If gas generation rates are fairly constant (no big increases and less than G1 limits above), what actions need to be taken if a transformer exceeds the L1 limits? Begin to pay more attention to that transformer, just as with a mineral oil transformer. It may be necessary to shorten the DGA sampling interval, reduce loading, check transformer cooling, get some outside advice, etc. As with mineral oil transformers, age exerts a big influence in accumulated gas. Be much more concerned if a 3-year old transformer has exceeded the L1 limits than if a 30-year old transformer exceeds the limits. However, if G1 generation rates are exceeded in either an old or new transformer, the level of concern should be stepped up.

If accumulated gas exceeds the L2 limit, it may be wise to plan to have the transformer de-gassed. Examine the physical tests in the DGAs and compare them to the Doble/IEEE table (table 23) (*Reference Book on Insulating Liquids and Gasses*) [25]. The oil should be treated in whatever manner is appropriate, if these limits are exceeded.

If both L1 limits and G1 limits are exceeded, be **more concerned**. Reduce sampling intervals, get outside advice, reduce loading, check transformer cooling and oil levels, etc. **If G2 generation limits are exceeded, be extremely concerned.** It will not be long before L3 limits are exceeded, and consider removing the transformer from service for testing, repair, or replacement.

If acetylene is being generated, the transformer should be taken out of service. However, as with mineral oil transformers, a one-time nearby lightning strike or through fault can cause a "one-time" generation of acetylene. If you notice acetylene in the DGA, immediately take another sample. **If the amount of acetylene is increasing, an active electrical arc is present within the transformer. It should be taken out of service.**

With critical silicone (or mineral oil-filled transformer), such as a single station service transformer or excitation transformer, a spare should be available at another facility. If there are no other possible spares, consider beginning the budget process for purchasing a spare transformer.

8.5 Physical Test Limits

Table 23 lists test limits for service-aged silicone filled transformer oil. If any of these limits are exceeded, treat the oil in whatever manner is appropriate to return the oil to serviceable condition.

Table 23 – Doble and IEEE Physical Test Limits for Service-Aged Silicone Fluid

Test	Acceptable Limits	Unacceptable Values Indicate	ASTM Test Method
Visual	Clear free of particles	Particulates, free water	D 1524 D 2129
Dielectric breakdown voltage	30 kV	Particulates, dissolved water	D 877
Water content maximum	70 ppm (Doble) 100 ppm (IEEE)	Dissolved water contamination	D 1533
Power factor maximum at 25 °C	0.2	Polar/ionic contamination	D 924
Viscosity at 25 °C, cSt	47.5–52.5	Fluid degradation contamination	D 44
Acid neutralization number maximum, mg KOH/gm	0.1 (Doble) 0.2 (IEEE)	Degradation of cellulose or contamination	D 974

Note: If only one number appears, both Doble and IEEE have the same limit.

If the above limits are exceeded in the DGA, the silicone oil should be filtered, dried, or treated to correct the specific problem.

9. Transformer Testing

Transformer testing falls into three broad categories: factory testing when the transformer is new or has been refurbished, acceptance testing upon delivery, and field testing for maintenance and diagnostic purposes. Some tests at the factory are common to most power transformers, but many of the factory tests are transformer-specific. Figure 68 (shown later in this manual) lists several tests. This test chart has been adapted from IEEE 62-1995 reference [20]. Not all of the listed tests are performed at the factory, and not all of them are performed in the field. Each transformer and each situation is different, requiring its own unique approach and tests.

9.1 DC Winding Resistance Measurement

CAUTION:

Do not attempt to run an excitation current test immediately after any direct current (dc) test. Energizing with dc will leave a residual magnetism in the core and will ruin the results of the excitation current test.

If generation of ethylene, ethane, and perhaps methane in the DGAs indicates a poor connection, winding resistances should be checked. Turns ratio, sweep frequency response analysis (SFRA), Doble tests, or relay operations may give indications that dc testing is warranted. Winding resistances are tested in the field to check for loose connections on bushings or tap changers, broken strands, and high contact resistance in tap changers. Results are compared to other phases in wye-connected transformers or between pairs of terminals on a delta-connected winding to determine if a resistance is too high. Resistances can also be compared to the original factory measurements or to sister transformers. Agreement within 5% for any of the above comparisons is considered satisfactory. If winding resistances are to be compared to factory values, resistance measurements will have to be converted to the reference temperature used at the factory (usually 75 °C). To do this, use the following formula:

$$Rs = Rm\frac{Ts + Tk}{Tm + Tk}$$

Rs =   Resistance at the factory reference temperature (found in the transformer manual)

Rm =   Resistance you actually measured

Ts =   Factory reference temperature (usually 75 °C)

Tm =   Temperature at which you took the measurements

Tk =   Constant for the particular metal the winding is made from: 234.5 °C for copper, 225 °C for aluminum

It is very difficult to determine actual winding temperature in the field; and, normally, this is not needed. The above temperature corrections are necessary only if resistance is to be compared to factory values. Normally, phase resistances are compared to each other or to sister transformers at the same temperature, and actual winding temperatures and corrections are not needed.

Compare winding resistances to factory values; change in these values can reveal serious problems. A suggested method to obtain an accurate temperature is outlined below. If a transformer has just been de-energized for testing, the winding will be cooler on the bottom than the top, and the winding hot spot will be hotter than the top oil temperature. The average winding temperature is needed, and it is important to get the temperature as accurate as possible for comparisons.

The most accurate method is to allow the transformer to sit de-energized until temperatures are equalized. This test can reveal serious problems, so it's worth the effort.

Winding resistances are measured using a Wheatstone Bridge for values of 1 ohm or above and using a micro-ohmmeter or Kelvin Bridge for values under 1 ohm. A multi-amp (now AVO) makes a good instrument for these measurements, which is quick and easy to use. Take readings from the top of each bushing to neutral for wye-connected windings and across each pair of bushings for delta-connected windings. If the neutral bushing is not available on wye-connected windings, take each one to ground (if the neutral is grounded) or take readings between pairs of bushings as if it were a delta winding. Be consistent each time so that a proper comparison can be made. The tap changer can also be changed from contact to contact, and the contact resistance can be checked. Make sure to take the first test with the tap changer "as found." Keep accurate records and connection diagrams so that later measurements can be compared.

9.2 Core Insulation Resistance and Inadvertent Core Ground Test (Megger®)

CAUTION:

Do not attempt to run excitation or SFRA tests on a transformer immediately after using dc test equipment. Residual magnetism will remain in the core and ruin the excitation current and SFRA test results.

Core insulation resistance and core ground test is used if an unintentional core ground is suspected; this may be indicated by the DGA. Key gases to look for are ethane and/or ethylene and possibly methane. These gases may also be present if there is a poor connection at the bottom of a bushing or a bad tap changer contact.

Therefore, this test is only necessary if the winding resistance test (section 9.1) shows all connections and tap changer contacts in good condition.

The intentional core ground must be disconnected. This may be difficult, and some oil may have to be drained to accomplish this. On some transformers, core grounds are brought outside through insulated bushings and are easily accessed. A standard dc Megger® (1,000-volt Megger® is recommended) is then attached between the core ground lead (or the top of the core itself and the tank [ground]). The Megger® is used to place a dc voltage between these points, and the resistance is measured. A new transformer should read greater than 1,000 megohms. A service-aged transformer should read greater than 100 megohms. Ten to one-hundred megohms is indicative of deteriorating insulation between the core and ground. Less than 10 megohms is sufficient to cause destructive circulating currents and must be further investigated [20]. A solid, unintentional core ground may read zero ohms; this, of course, causes destructive circulating currents and must be corrected before energization.

Some limited success has been obtained in "burning off" unintentional core grounds using a dc or ac current source. This is a risky operation, and the current may cause additional damage. The current source is normally limited to a maximum of 40 to 50 amps and should be increased slowly to use as little current as possible to accomplish the task. This should only be used as a last resort and then only with consultation from the manufacturer, if possible, and with others experienced in this task.

CAUTION:

This will generate gases which will be dissolved in the oil and will show up in the DGA! Take a sample for DGA with in 72 hours after burning off the unintentional core ground and compare this DGA with the prior one to determine what gases were created by this task.

9.3 Doble Tests on Insulation

Doble testing is important to determine the condition of a transformer, because it can detect winding and bushing insulation integrity and problems in the winding and core. Doble tests are conducted in the field on de-energized transformers using special test equipment. Generally, a Doble M-4000 test set is used along with accompanying software. The software automatically performs analysis of test results and responds with a four letter code: G = Good, I = Investigate, D = Deteriorated, and B = Bad. These codes refer to insulation quality. If a "D" or "B" code is encountered, the insulation should be re-tested, carefully investigated, and the problem explained before re-energizing. Other tests may have to be performed; and, perhaps, an internal inspection should be considered before the unit is re-energized. The Doble Company should be consulted, along with the transformer manufacturer, and other transformer experts. If the problem is severe, the unit may have to be taken out of service.

9.3.1 Insulation Power Factor Test

The purpose of this test is to determine the state of dryness of the windings and insulation system and to determine a power factor for the overall insulation, including bushings, oil, and windings. It is a measure of the ratio of the power (I^2R) losses to the voltamperes applied during the test. The power factor obtained is a measure of watts lost in the total transformer insulation system, including the bushings. The power factor should not exceed 0.5% at 20 C. Temperature correction of test results can be performed automatically on the Doble test set. The watts lost should not exceed one-half of one percent of the total power input (voltamperes) from the test. The values obtained at each test are compared to previous tests and baseline factory tests, and a trend can be established as the insulation system ages.

9.3.2 Capacitance Test

This test measures and records the capacitance (including bushings) between the high and low-voltage windings, between the high-voltage winding and the tank (ground), and between the low-voltage winding and the tank (ground). Changes in these values as the transformer ages and events occur, such as nearby lightning strikes or through faults, indicate winding deformation and structural problems, such as displaced wedging and winding support.

9.3.3 Excitation Current Test

CAUTION:

Perform the excitation test before any dc tests. Excitation current tests should never be conducted after a dc test has been performed on the transformer. Results will be incorrect because of residual magnetism of the core left from the dc tests.

The purpose of this test is to detect short-circuited turns, poor electrical connections, core de-laminations, core lamination shorts, tap changer problems, and other possible core and winding problems. On three-phase transformers, results are also compared between phases. This test measures current needed to magnetize the core and generate the magnetic field in the windings. Doble software only gives two indications on this test: "G" for good and "Q" for questionable. On a three-phase, wye/delta or delta/wye transformer test, the excitation current pattern will be two phases higher than the remaining phase. Compare the two higher currents only. If the excitation current is less than 50 milliamps (mA), the difference between the two higher currents should be less than 10%. If the excitation current is more than 50 mA, the difference should be less than 5%. In general, if there is an internal problem, these differences will be greater. When this happens, other tests should also show abnormalities, and an internal inspection should be considered. The results, as with all others, should be compared with factory and prior field tests.

9.3.4 Bushing Tests

For bushings that have a potential tap, both the capacitance between the top of the bushing and the bottom tap (normally called C1), and the capacitance between the tap and ground (normally called C2) are measured. To determine bushing losses, power factor tests are also performed. C2 capacitance is much greater than C1. Bushings without a potential tap are normally tested from the bushing top conductor to ground and "hot collar" tests. These test results are compared with factory tests and/or prior tests to determine deterioration. About 90% of bushing failures may be attributed to moisture ingress, evidenced by an increasing power factor from Doble testing on a scheduled basis.

9.3.5 Percent Impedance/Leakage Reactance Test

This is normally an acceptance test to see that nameplate percent impedance agrees with the measured percent impedance when the transformer arrives onsite. Normally, a 3% difference is considered acceptable. However, after the initial benchmark test, the percent impedance should not vary more than 2% from benchmark. As the transformer ages or suffers events such as through faults, nearby lightning strikes, and other surges, this test is used in the field to detect winding deformation. Winding deformation can lead to immediate transformer failure after a severe through fault, or a small deformation can lead to a failure years later.

Percent impedance/leakage reactance testing is performed by short circuiting the low-voltage winding and applying a test voltage to the high-voltage winding. Reluctance is resistance to lines of magnetic flux. Reluctance to the magnetic flux is very high in spaces between the high- and low-voltage windings and spaces between the windings and core. Reluctance is very low through the magnetic core so that the vast majority of total reluctance is in the spaces. When winding movement (distortion) occurs, these spaces change. Therefore, the reluctance changes, resulting in a change in the measured leakage reactance. Changes in leakage reactance and in capacitance tests (explained in section 9.3.2) serve as an excellent indicator of winding movement and structural problems (displaced wedging, etc.). This test does not replace excitation current tests or capacitance tests but complement them, and they are used together. The excitation current test relies on reluctance of the core while the leakage reactance test relies on reluctance of the spaces. See *Doble's Leakage Reactance Instrument Users Guide, and IEEE®, Guide for Diagnostic Field Testing of Electric Power Apparatus-Part 1: Oil-Filled Power Transformers, Regulators, and Reactors* (IEEE 62-1995™ [20]).

9.3.6 Sweep Frequency Response Analysis Tests

CAUTION:

Do not attempt to run a sweep frequency response analysis test immediately after any dc test. Energizing with dc will leave a residual magnetism in the core and will offset the results of the SFRA test.

These tests show, in trace form, the winding transfer function of the transformer and are valuable to determine if any damage has occurred during shipping or during a through fault. Core grounds, core displacement, and other core and winding problems can be revealed by this test.

These tests should be conducted before and after the transformer has been moved or after experiencing a through fault. Results should be compared to baseline tests performed at the factory or as soon as possible after receiving the transformer. If the SFRA tests cannot be performed at the factory, they should be conducted as an acceptance test before energizing a new or rebuilt transformer to establish a baseline. A baseline should be established for older inservice transformers during a normal Doble test cycle. If possible, one should use the same test equipment for baseline and following tests, or the results may not be comparable.

For a delta/wye transformer, a test voltage of variable frequency (normally 20 hertz [Hz] to 2 megahertz [MHz]) is placed across each phase of the high-voltage winding. With this set of tests, low-voltage windings are isolated with no connections on any of the bushings. An additional set of tests is performed by short circuiting all the low-voltage windings and again placing the test voltage on each phase of the high-voltage winding. A third set of tests is made by isolating the high-voltage winding and placing the test voltage across each low-voltage winding. See the *M5100 SFRA Instrument Users Guide*, Doble Engineering Company [36] for connection details.

Figure 62 is a picture of test traces on a new three-phase Reclamation transformer. The top three traces were taken on the low-voltage side: X1-X3, X2-X1, and X3-X2. The two outside windings (A and C phases) have the same general shape with a "W" at the lowest point of the trace, while the inside winding (B phase) has a "V" at the bottom. The high-voltage traces (lower three) have the same characteristics. Note that, in both high- and low-voltage tests, the traces fall almost

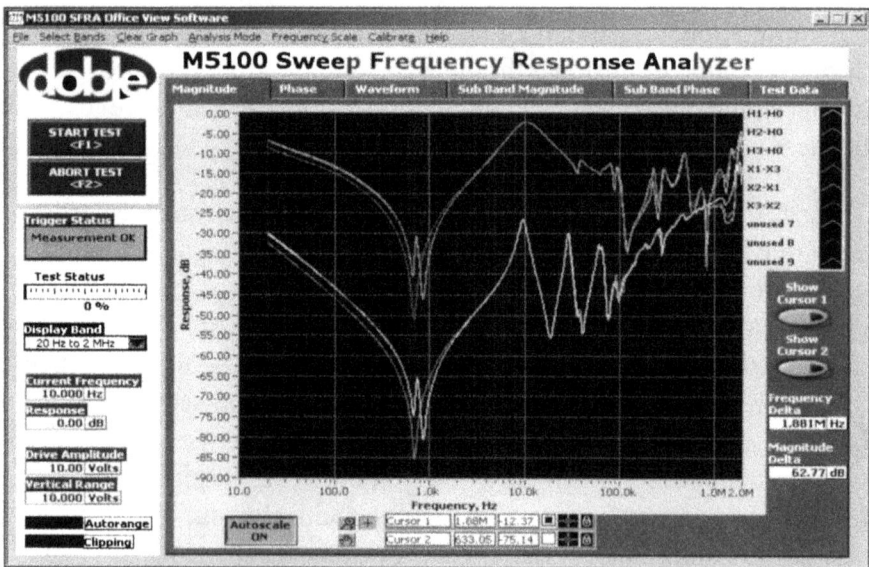

Figure 62 – SFRA Test Traces of a New Transformer.

perfectly on top of each other for the outside windings (A and C), while the inside winding (B phase), is slightly displaced to the left. These are characteristic traces of a three-phase transformer in good condition; these traces will be the baseline for future tests on this transformer.

By comparing future traces with baseline traces, the following can be noted. In general, the traces will change shape and be distorted in the low frequency range (under 5,000 Hz) if there is a core problem. The traces will be distorted and change shape in higher frequencies (above 10,000 Hz) if there is a winding problem. Changes of less than 3 decibels (dB) compared to baseline traces are normal and within tolerances. From 5 Hz to 2 kilohertz (kHz), changes of + or − 3 dB (or more) can indicate shorted turns, open circuit, residual magnetism, or core movement. From 50 Hz to 20 kHz +/- 3 dB (or more), a change from baseline can indicate bulk movement of windings relative to each other. From 500 Hz to 2 MHz, changes of +/- 3 dB (or more) can indicate deformation within a winding. From 25 Hz to 10 MHz, changes of +/- 3 dB (or more) can indicate problems with winding leads and/or test leads placement. The above diagnostics come from the Euro-Doble Client Committee after much testing experience and analysis. Note that there is a great deal of overlap in frequencies, which can mean more than one diagnosis.

Figure 63 shows traces of a transformer with a problem. This transformer is not Reclamation-owned, and test results are used for illustration only. The traces have the same general positions on the graph as the good transformer. The lower traces are high-voltage winding tests, while the upper traces are the low-voltage winding tests. Note in the higher frequencies of the low-voltage traces that "A" phase (X1-X0 green trace) is displaced from the other two phases more than 3 dB from about 4 kHz to about 50 kHz. With a healthy transformer, these would fall almost on top of each other as the other two phases do. Also notice that "A" phase (H1-H3 Lsh) is displaced in the test with the low-voltage winding shorted. There is an obvious problem with "A" phase on the low-voltage side. After opening the

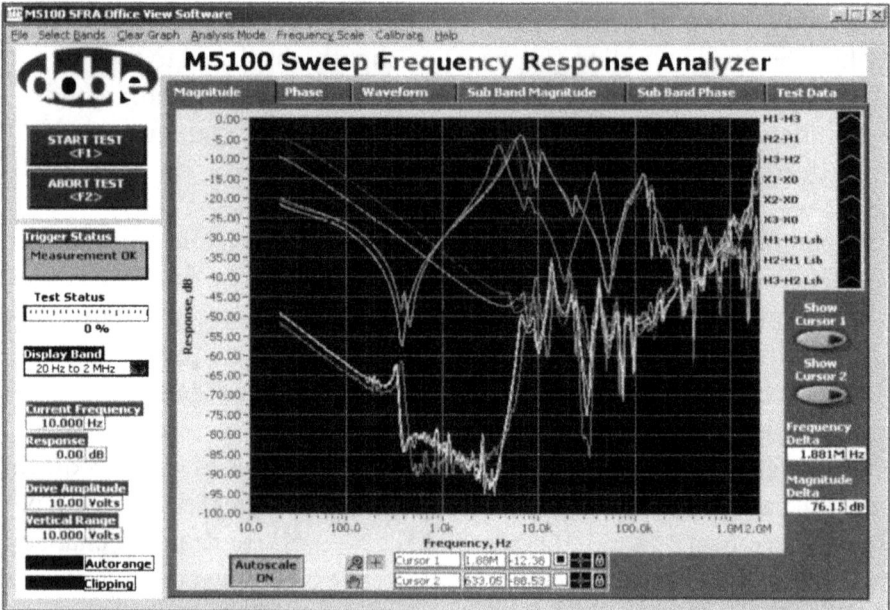

Figure 63 – SFRA Test Traces of a Defective New Transformer.

transformer, it was found that the "A" phase winding lead had burned off near the winding connection and re-welded itself on the winding at a different location, effectively shorting out a few turns. The transformer was still working, but hot metal gases (ethylene, ethane, methane) were actively generating and showing up in the DGA. Although other tests could have revealed this problem, SFRA showed the problem was with "A" phase and, therefore, where to concentrate the internal inspection.

9.4 Visual Inspection

9.4.1 Background
Visual inspection of the transformer exterior reveals important condition information. For example, valves positioned incorrectly, plugged radiators, stuck temperature indicator and level gauges, and noisy oil pumps or fans. Oil leaks can often be seen which indicate a

potential for oil contamination, loss of insulation, or environmental problems. Physical inspection requires staff experienced in these techniques. There are several required physical inspections that will not be covered here because they were addressed previously. Be sure to inspect the following as explained in each respective section: winding temperature indicators (section 4.3), pressure relief devices (section 4.5), sudden pressure relay (section 4.6), conservator bladder (section 4.9.2.9), conservator breather (section 4.9.2.9.2), and bladder failure relay (section 4.9.2.9.3).

9.4.2 Oil Leaks

Check the entire transformer for oil leaks. Leaks develop due to gaskets wearing out, ultraviolet exposure, taking a "set," or from expansion and contraction, especially after transformers have cooled, due to thermal shrinkage of gaskets and flanges. Many leaks can be repaired by applying an epoxy or other patch. Flange leaks may be stopped with these methods, using rubberized epoxy forced into the flange under pressure. Very small leaks in welds and tanks may be stopped by peening with a ball-peen hammer, cleaning with the proper solvent, and applying a "patch" of the correct epoxy. Experienced leak mitigation contractors whose work is guaranteed may also be employed. Some leaks may have to be welded. Welding may be performed with oil in the transformer if an experienced, qualified, and knowledgeable welder is available. If welding with oil in the tank is the method chosen, oil samples must be taken for DGA, both before and after welding. Welding may cause gases to appear in the DGA and it must be determined what gases are attributed to welding and which ones to transformer operation. See also EPRI's *Guidelines for the Life Extension of Substations*, 2000 and 2002 update [5] section on "Transformers Leak Mitigation" that gives several materials and applications to stop transformer oil leaks. Copies of this document are available at the Technical Service Center.

9.4.3 Oil Pumps

If the transformer has oil pumps, check the flow indicators and pump isolation valves to ensure that oil is circulating properly. Pump motor(s) may also have reversed rotation, and flow indicators may still show that oil is flowing. To ensure motors are turning in the proper direction, use an ammeter to check the motor current. Compare results with the full-load-current indicated on the motor nameplate. If the motor is reversed, the current will be much less than the nameplate full-load-current. Check oil pumps with a vibration analyzer if they develop unusual noises. Have the DGA lab check for dissolved metals in the oil and run a metal particle count for metals, if the bearings are suspect. This should be performed as soon as a bearing becomes suspect; bad oil pump bearings can put enough metal particles into the oil to threaten transformer insulation and cause flashover inside the tank, resulting in an explosive catastrophic failure of the transformer tank.

9.4.4 Fans and Radiators

Inspect all isolation valves at the tops and bottoms of radiators to ensure they are open. Inspect cooling fans and radiators for cleanliness and fans for proper rotation. Check for dirty or damaged fan blades or partially blocked radiators. Fans are much more efficient if the blades are clean and rotating in cool air. Normally, fans blow cool air through the radiators; they should not be pulling air through. Check to see if fans are reversed electrically (i.e., pulling air first through the radiators and then through the fan blades). This means that the blades are rotating in warm air after it passes through the radiator, which is much less efficient. Place a hand on the radiator opposite the fans; air should be coming out of the radiator against your hand. Watch the blades as they rotate slowly when they are starting or stopping to determine which way they should be rotating and correct the rotation if necessary. See IEEE 62-1995™ [20]. Also inspect radiators and fans with an infrared camera, see section 9.4.6, "Infrared Temperature Analysis."

9.4.5 Age

Transformer age is an important factor to consider when identifying candidates for replacement or rehabilitation. Age is one indicator of remaining life and upgrade potential to current state-of-the art materials. During transformer life, structural strength and insulating properties of materials used for support and electrical insulation (especially paper) deteriorate. Although actual service life varies widely depending on the manufacturer, design, quality of assembly, materials used, maintenance, and operating conditions, the expected life of a transformer is about 40 years (Replacements, July 1995, Bureau of Reclamation and Western Area Power Administration [35]).

9.4.6 Infrared Temperature Analysis
Infrared analysis should be conducted annually while equipment is energized and under full load, if possible. IR analysis should also be conducted after any maintenance or testing to see if connections that were broken were re-made properly. Also, if IR is performed during factory heat run, the results can be used as a baseline for later comparison.

9.4.7 IR for Transformer Tanks
Unusually high external temperatures or unusual thermal patterns of transformer tanks indicate problems inside the transformer, such as low oil level, circulating stray currents, blocked cooling, loose shields, tap changer problems, etc. Infrared scanning and analysis is required annually for trending purposes by National Fire Protection Association 70B, *Recommended Practice for Electrical Equipment Maintenance* [8] and *FIST Volume 4-1B* [37]. See also *FIST Volume 4-13* [38] and IEEE 62-1995™ [20]. Abnormally high temperatures can damage or destroy transformer insulation and, thus, reduce life expectancy. Thermal patterns of transformer tanks and radiators should be cooler at the bottom and gradually warmer ascending to the top. See figure 64 for a normal pattern; the red spot at the top is normal showing a "hot spot" top of B phase, about 110 degrees Fahrenheit (°F). Any departure from this pattern means a probable problem which must be investigated. An IR inspection can find over-heating conditions or incorrect thermal patterns. IR scanning and analysis requires trained staff, experienced in these techniques.

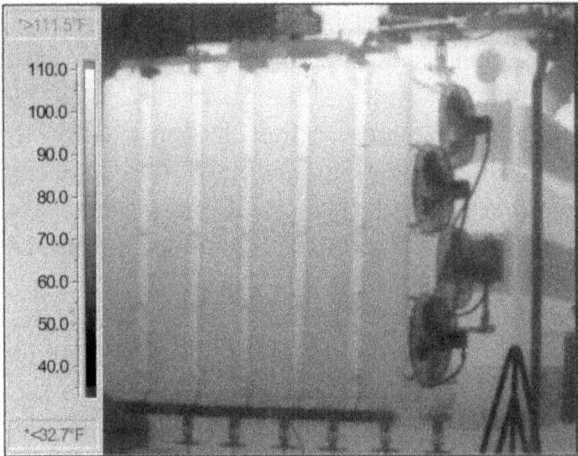

Figure 64 – Normal Transformer IR Pattern.

9.4.8 IR for Surge Arresters

Surge arresters should be included when infrared scanning energized transformers. Look for unusual thermal patterns on the surface of lightning arresters (see the arrester IR image in figure 65). Note that the yellow in the top right of the image is a reflection not associated with the arrester. A temperature profile of the arrester is shown as black lines. Note the hot spot (yellow) about a third of the way down from the top. This indicates that immediate de-energization and replacement must be undertaken. Catastrophic failure is imminent which can destroy nearby equipment and be hazardous to workers. Also compare thermal patterns to sister units or earlier scans of the same arrester. Scan all high-voltage connections and compare them to nearby connections for unusual temperatures.

9.4.9 IR for Bushings

IR scans of bushings can show low oil levels which would call for immediate de-energization and replacement. This generally means that the seal in the bushing bottom has failed, leaking oil into the transformer. The top seal has probably also failed, allowing air and up the bushing. Another reason a bushing can exhibit a high oil level

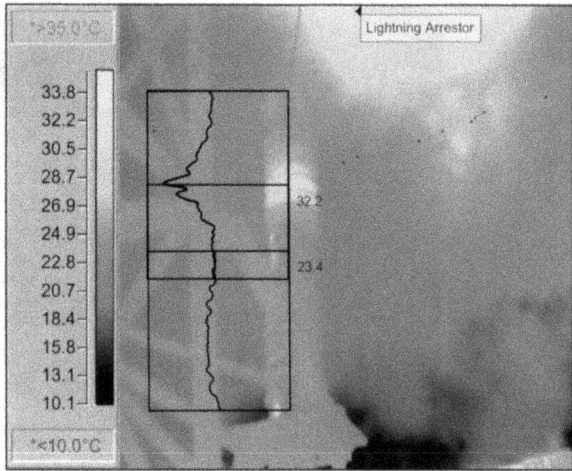

Figure 65 – IR Image of Defective Arrester.

is that the top seal is leaking, allowing water to enter. The water migrates to the bushing bottom, displacing the oil upward. Remember, over 90% of bushing failures are attributed to water entrance through the top seal. Bushings commonly fail catastrophically, many times destroying the host transformer or breaker and nearby equipment and causing hazards to workers. Figure 66 shows low-oil level in a high-voltage transformer bushing. Compare previous IR scans of the same bushing with the current scan. Doble hot-collar testing possibly may show this problem. However, Doble tests are run infrequently, and the transformer has to be out of service, under clearance, and both primary and secondary conductors removed, while an IR scan easily can be performed at any time.

9.4.10 IR for Radiators and Cooling Systems
Examine radiators with an IR camera and compare them with each other. A cool radiator or segment indicates that a valve is closed or the radiator or segment is plugged. The IR image (figure 67) shows that the cold left radiator section is valved off or plugged. If visual inspection shows the valves are open, the radiator or segment must be

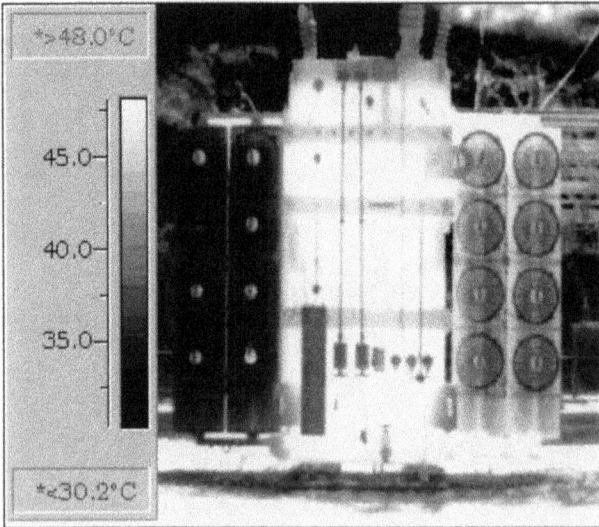

Figure 66 – IR Image Showing Blocked Radiators.

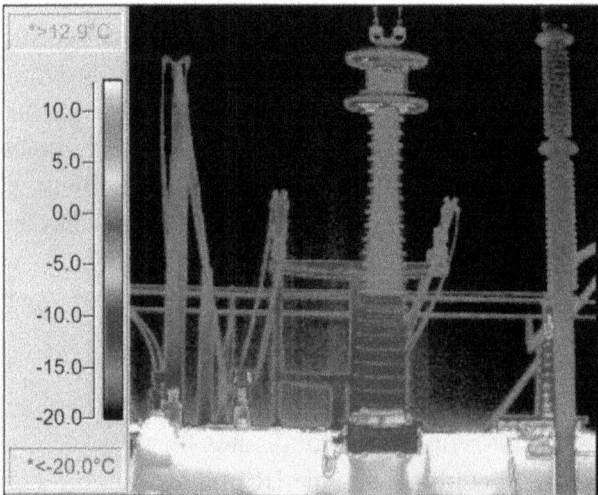

Figure 67 – IR Image of Defective Bushing.

isolated, drained, removed, and the blockage cleared. Do not allow a transformer to operate with reduced cooling which drastically shortens transformer life. Remember, an increased operating temperature of only 8 to 10 °C will reduce transformer life by one-half. IR scan all cooling systems, including heat exchangers, fans, pumps, motors, etc. Check inside control panels for overloaded wiring, loose connections, and overheated relays. Look for unusual thermal patterns and compare similar equipment.

9.4.11 Corona Scope Scan
With the transformer energized, scan the bushings and surge arresters and all high-voltage connections for unusual corona patterns. Corona should be visible only at the top of bushings and arresters, and corona at connections should be similar to sister connections. As a bushing deteriorates due to physical defects, the corona pattern will grow progressively larger. When the corona pattern reaches a grounded surface (i.e., the tank or structure), a flashover will occur, destroying the bushing or arrester and, perhaps, the transformer. The corona scope will reveal this problem long before a flashover.

9.5 Ultrasonic and Sonic Fault Detection

9.5.1 Background
This test should be performed when hydrogen is increasing markedly in the DGA. High hydrogen generation indicates partial discharge occurring inside the transformer. Other gases such as methane, ethane, and ethylene may also be increasing. Acetylene may also be present, if arcing is occurring, and may also be increasing. Ultrasonic contact (in contact with the tank) fault detection can detect partial discharge (corona) and full discharge (arcing) inside the transformer. This test can also detect loose parts inside the transformer. Partial discharges emit energy in the order of 20 kHz to 200 kHz. These frequencies are above levels that can be audibly detected. The test equipment receives the signals and converts them electronically into audible signals. Headphones are provided to eliminate spurious noise from the

powerplant and other sources. The equipment logs data for future reference. A baseline test should be conducted and compared with future test data. This test method has some limitations; if a partial discharge is located deep within the windings, external detectors may not be sensitive enough to detect and locate the problem. However, partial discharges most often occur near the top of the transformer in areas of high-voltage stress which can readily be located by this method. These defects can sometimes be easily remedied, extending transformer service life.

9.5.2 Process

Magnetic piezoelectric crystal transducers, sized and tuned to the appropriate frequencies, are placed on the outside of the tank, and signals are recorded. If discharges are detected, the location is triangulated so that, during an internal inspection, the inspector will know the general area to search for a problem. Likewise, sonic (audible ranges) fault detection can find mechanical problems, such as noisy bearings in pumps or fans, nitrogen leaks, loose shields, or other loose parts inside the transformer tank, etc. See EPRI's *Guidelines for the Life Extension of Substations* [5], section 3. See also IEEE 62-1995™ [20], section 6.1.8.4.

9.6 Vibration Analysis

9.6.1 Background

Vibration analysis by itself cannot predict many faults associated with transformers, but it is another useful tool to help determine transformer condition. Vibration can result from loose transformer core segments, loose windings, shield problems, loose parts, or bad bearings on oil cooling pumps or fans. Exercise extreme care in evaluating the source of vibration. Many times, a loose panel cover, door, or bolts/screws lying in control panels, or loose on the outside have been misdiagnosed as problems inside the tank. There are several instruments available from various manufacturers, and the technology is advancing quickly. Every transformer is different; therefore, to detect this, baseline vibration tests should be run and data recorded for comparison with future tests.

9.6.2 Process

For a normal transformer in good condition, vibration data is normally two times line frequency (120 Hz) and also appears as multiples of two times line frequency; that is, four times 60 (240 Hz), six times 60 (360 Hz), etc. The 120 Hz is always the largest and has an amplitude of less than 0.5 inch per second (ips) and greater than 0.1 ips. The next peak of interest is the four times line frequency or 240 Hz. The amplitude of this peak should not exceed 0.5 ips. None of the remaining harmonic peaks should exceed 0.15 ips in amplitude. See *EPRI's Proceedings: Substation Equipment Diagnostics, Conference IX* [24] section on "Vibration Analysis."

9.7 Turns Ratio Test

9.7.1 Background

This test only needs to be performed if a problem is suspected from the DGA, Doble testing, or relay operation. The turns ratio test detects shorted turns, which indicate insulation failure by determining if the correct turns ratio exists. Shorted turns may result from short circuits or dielectric (insulation) failures.

9.7.2 Process

Measurements are taken by applying a known low voltage across one winding and measuring the induced-voltage on the corresponding winding. The low voltage is normally applied across a high-voltage winding so that the induced-voltage is lower, reducing hazards while performing the test. The voltage ratio obtained by the test is compared to the nameplate voltage ratio. The ratio obtained from the field test should agree with the factory within 0.5%. New transformers of good quality normally compare to the nameplate within 0.1%.

For three-phase delta/wye or wye/delta connected transformers, a three-phase equivalency test should be performed. The test is

195

performed and calculated across corresponding single windings. Look at the nameplate phasor diagram to find out what winding on the primary corresponds to a particular winding on the secondary. Calculate the ratio of each three-phase winding based on the line to neutral voltage of the wye winding. Divide the line-to-line winding voltage by 1.732 to obtain the correct line-to-neutral voltage. Check the tap changer position to make sure it is set at the position where the nameplate voltage is based. Otherwise, the turns ratio test information cannot be compared with the nameplate. Nameplate information for Reclamation transformers is normally based on the tap three position of the tap changer. See the manufacturer's instruction manual for the specific turns ratio tester for details. See IEEE 62-1995™ [20].

9.8 Estimate of Paper Deterioration (Online)

NOTE:

The two methods below should be used together.

9.8.1 CO_2 and CO Accumulated Total Gas Values
IEEE Standard C57.104™ *Guide for the Interpretation of Gases Generated in Oil-Immersed Transformers* [12] gives status conditions, based on accumulated values of CO_2 and CO. Accumulated dissolved gas levels provide four status conditions: normal operation, modest concern (investigate), major concern (more investigation), and imminent risk (nearing failure). The CO_2 and CO levels in ppm for each status are given in table 24.

9.8.2 CO_2/CO Ratio
The CO_2/CO ratio should also be analyzed in conjunction with the CO_2 and CO accumulated values. For analyzing the CO_2/CO ratio, see section 6.1.10.

Table 24 – Paper Status Conditions Using CO_2 and CO

Condition 1	Normal	0 – 2,500	0 – 350
Condition 2	Modest Concern	2,500 – 4,000	351 –570
Condition 3	Major Concern	4,001 – 10,000	571 – 1,400
Condition 4	Imminent Risk	>10,000	>1,499

CAUTION:

The status from table 24 should be at least in Condition 2 or 3 from one or both gases before a detailed investigation is begun. There is no need to look at the ratios from section 6.1.10 unless a substantial amount of these gases have already been generated. If the transformer is relatively new, CO_2 and other atmospheric gases (N_2, O_2, and even some CO) may be migrating out of the paper into the oil because the paper was stored in air prior to transformer assembly. If the paper was stored in a polluted city atmosphere, a considerable amount of CO may show up in the DGA. This may look like the transformer has a problem and is generating a lot of CO. However, if the transformer has a real problem, H_2 and perhaps other heat gases (CH_4, C_2H_6, C_2H_4) should also be increasing.

9.9 Estimate of Paper Deterioration (Offline During Internal Inspection)

9.9.1 Degree of Polymerization (DP)
Do not open a transformer for the sole purpose of doing this test. Perform this test only if the unit is being opened for other reasons. See the reasons for an internal inspection in section 9.9.2.

9.9.1.1 Background – One of the most dependable means of determining paper deterioration and remaining life is the DP test of the cellulose. The cellulose molecule is made up of a long chain of glucose rings which form the mechanical strength of the molecule and the paper. DP is the average number of these rings in the molecule.

197

As paper ages or deteriorates from heat, acids, oxygen, and water, the number of these rings decreases. When the insulation is new, the DP is typically between 1,000 and 1,400. As paper deteriorates, bonds between the rings begin to break. When the DP reaches around 200, the insulation has reached the end of life. All mechanical strength of the insulation has been lost; the transformer must be replaced.

   9.9.1.2 Process – When doing an internal inspection, or if the transformer is opened and oil is fully or partially drained for any reason on a service-aged transformer, perform a DP test. Remove a sample of the paper insulation about 1 centimeter square from a convenient location near the top of the center phase with a pair of tweezers. In general, in a three-phase transformer, the hottest most thermally aged paper will be at the top of the center phase. If it is not possible to take a sample from the center phase, take a sample from the top of one of the other phases. Send this sample to an oil testing laboratory for the DP test. Analyze results of the DP test with table 25 taken from EPRI's *Guidelines for the Life Extension of Substations,* 2002 Update, chapter 3 [5]. Table 25 has been developed by EPRI to estimate remaining life.

Table 25 – DP Values for Estimating Remaining Paper Life

New insulation	1,000 DP to 1,400 DP
60% to 66 % life remaining	500 DP
30% life remaining	300 DP
0 life remaining	200 DP

9.9.2 Internal Inspection

   9.9.2.1 Background – If an internal inspection is absolutely necessary, it must be completed by an experienced person who knows exactly what to look for and where to look. Many times, more damage is done by opening a transformer and doing an internal inspection than what is gained. There are very few reasons for an internal inspection; some are shown below:

♦ Extensive testing shows serious problems.

♦ Unexplained relay operation takes the transformer offline, and testing is inconclusive.

♦ Acetylene is being generated in the DGA (indicates active internal arcing).

♦ Ethylene and ethane are being generated in sufficient quantities to cause grave concern. This generally indicates a bad connection on a bushing bottom or tap changer, circulating currents, additional core ground, or static discharges.

♦ A core ground must be repaired, or an additional core ground has developed which must be removed.

♦ Vibration and ultrasonic analysis indicate loose windings that are generating gases from heat caused by friction of the vibrating coils; loose wedges must be located and replaced.

♦ CO_2/CO ratio are very low (around 2 or 3), indicating severe paper deterioration from overheating. Cooling must be checked carefully before opening the transformer.

♦ Furans are high (see section 7.6), indicating excessive aging rate; a DP test must be completed.

♦ The metal particle count is above 5,000 in 10 milliliters of oil taken specifically for detecting metal particle count. (See EPRI's *Guidelines for the Life Extension of Substations*, 2002 Update [5] for exact procedures for detecting and correcting metal particle contamination).

NOTE:

If a service-aged transformer is opened for any reason, a sample of the paper should be taken for DP analysis. If possible, take the paper sample from the top of the center phase winding because this will be near the hot spot. If it is not possible to get a paper sample from the top of B phase, take a sample from the top of one of the other windings.

9.9.2.2 Transformer Borescope – A new technology has been developed for internal transformer inspections using a specifically designed borescope. The borescope can be used with oil inside the transformer; core, windings, connections, etc., can be examined and photographed. If it is necessary to go inside the transformer for repairs, workers will possibly know exactly what is defective and exactly what must be done. This technology, used properly, can save generating time and repair dollars.

Currently, this technology is available only as a contracting service. However, plans are being made by the developing company to make the instrument available for sale.

9.10 Transformer Operating History

One of the most important transformer diagnostic tools is the operating history, including temperatures, overload history (especially through faults), relay operations, and nearby lightning strikes. Many times, a through fault or nearby lightning strike will generate acetylene or ethylene and other gases, and these will remain in future DGAs. Through faults (i.e., where fault current passes through the windings) subject a transformer to severe electrical and mechanical stresses which can degrade insulation and mechanical strength. Likewise, close-in lightning strikes subject the transformer to stresses. Excessive heating caused by sustained overloads or loss of cooling will degrade insulation. The damage may not be immediately evident but can show up years later. Through faults, overload history, and temperatures should be in operations records, and a review of these is required when diagnosing a transformer problem.

9.11 Transformer Diagnostics/Condition Assessment Summary

A summary of diagnostic techniques is given below in table 26.

Table 26 – Reclamation Transformer Condition Assessment Summary

Tests	Detects	Tools
Online Tests		
Dissolved Gas Analysis (Laboratory and Portable)	Measures dissolved gases: to detect, arcing, bad electrical contacts, hot spots, partial discharges, overheating of conductors, oil, tank, cellulose (paper insulation).	Portable Gas –In – Oil Analyzer with software[1] (e.g., Morgan Schaffer P200 and Transformer Oil Analyst software). Note: the P200 does not measure oxygen and nitrogen. Or send to an oil testing laboratory for DGA.
Oil Physical and Chemical Tests	Moisture, interfacial tension, acidity, furans, dissolved metals, and metal particle count (indicates pump problems). Analyzes paper and oil condition.	Requires laboratory analysis.
Physical Inspection – External	Oil leaks, broken parts, worn paint, defective support structure, malfunctioning temperature and level indicators, cooling problems, pump and radiator problems, bushing and lightning arrester porcelain cracks, etc.	Experienced staff, binoculars for bushing, and lightning arrester porcelain cracks and loose bolts.
Infrared Scan	Hot spots, localized heating, bad connections, circulating currents, blocked cooling, tap changer problems, bushing and lightning arrester problems.	Trained and experienced staff with thermographic camera and analysis software.[1]

Table 26 – Reclamation Transformer Condition Assessment Summary (continued)

Tests	Detects	Tools
Online Tests (continued)		
Ultrasonic and Sonic Contact Fault Detection	Internal partial discharge, arcing, sparking, pump impeller and bearing problems. Mechanical noises, loose parts (blocking, deflectors, etc.).	Fault detector with data logger.[1]
Ultrasonic Non-Contact and Contact Fault Detection	Nitrogen leaks, vacuum leaks, corona at bushings, pump mechanical and bearing problems, cooling fan problems.	Ultrasonic probe and meter (e.g., UltraProbe 2000).
Vibration Analysis	Internal core, shield problems. Loose parts vibration.	Vibration data logger.[1]
External Temperatures (Main Tank)	Temperature monitoring with changes in load and ambient temperature.	Portable temperature data loggers and software.[1]
Sound Level	Internal and external noises to compare to baseline and other vibration tests.	Sound level meter[1] (e.g., Quest 2400 Sound Meter).
Corona	Compare bushings and lightning arresters and all high-voltage connections with sister units.	Corona Scope,[1] Model CS-01-A.
Offline Tests		
Doble Power Factor	Loss of winding insulation integrity. Loss of bushing insulation integrity. Winding moisture.	Doble test equipment.
Excitation Current	Shorted turns in the windings.	Doble test equipment.
Turns Ratio	Shorted windings.	Turns ratio tester (e.g., Biddle 55005).
Leakage Reactance	Measures percent impedance, to be compared to nameplate after moving or through fault.	Doble test equipment.

Table 26 – Reclamation Transformer Condition Assessment Summary (continued)

Tests	Detects	Tools
Offline Tests (continued)		
Sweep Frequency Response Analysis	Structural problems, core and winding problems, movement of core and windings. Run this test before and after moving and after a through fault.	Doble M5100 sweep frequency response analyzer.
Across Winding Resistance Measurements	Broken strands, loose connections, bad contacts in tap changer.	Wheatstone Bridge (1 ohm and greater). Kelvin Bridge (less than 1 ohm).
Winding DC Resistance to Ground	Winding low resistance to ground (leakage current).	Megger®.[1]
Core to Ground Resistance[2]	Bad connection on intentional core ground or existence of unintentional grounds.	Megger®.[1]
Internal Inspections and Tests	Oil sludging, displaced winding or wedging, loose windings, bad connections, burned conductors.	Experienced staff, micro-ohm meter and/or Foster Miller borescope.
Degree of Polymerization	Insulation condition (life expectancy).	Laboratory analysis of paper sample.

[1] This equipment is available from the Hydroelectric Research and Technical Services Group.

[2] External test may be possible depending on transformer construction.

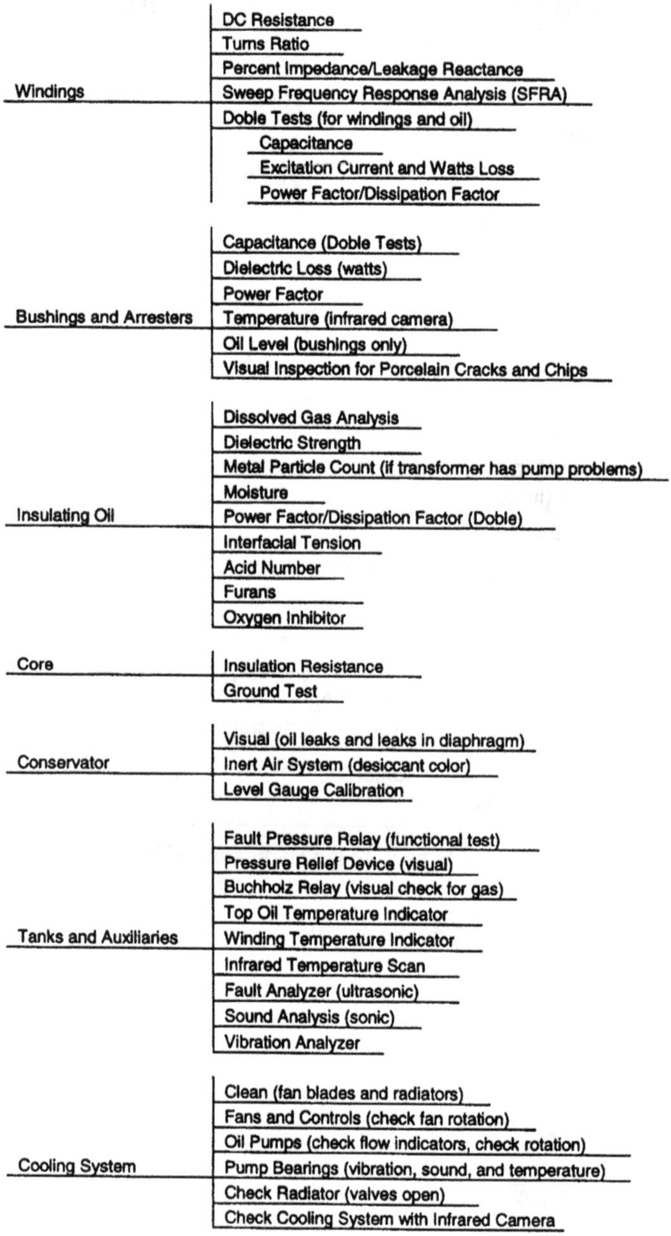

Windings
- DC Resistance
- Turns Ratio
- Percent Impedance/Leakage Reactance
- Sweep Frequency Response Analysis (SFRA)
- Doble Tests (for windings and oil)
  - Capacitance
  - Excitation Current and Watts Loss
  - Power Factor/Dissipation Factor

Bushings and Arresters
- Capacitance (Doble Tests)
- Dielectric Loss (watts)
- Power Factor
- Temperature (infrared camera)
- Oil Level (bushings only)
- Visual Inspection for Porcelain Cracks and Chips

Insulating Oil
- Dissolved Gas Analysis
- Dielectric Strength
- Metal Particle Count (if transformer has pump problems)
- Moisture
- Power Factor/Dissipation Factor (Doble)
- Interfacial Tension
- Acid Number
- Furans
- Oxygen Inhibitor

Core
- Insulation Resistance
- Ground Test

Conservator
- Visual (oil leaks and leaks in diaphragm)
- Inert Air System (desiccant color)
- Level Gauge Calibration

Tanks and Auxiliaries
- Fault Pressure Relay (functional test)
- Pressure Relief Device (visual)
- Buchholz Relay (visual check for gas)
- Top Oil Temperature Indicator
- Winding Temperature Indicator
- Infrared Temperature Scan
- Fault Analyzer (ultrasonic)
- Sound Analysis (sonic)
- Vibration Analyzer

Cooling System
- Clean (fan blades and radiators)
- Fans and Controls (check fan rotation)
- Oil Pumps (check flow indicators, check rotation)
- Pump Bearings (vibration, sound, and temperature)
- Check Radiator (valves open)
- Check Cooling System with Infrared Camera

Figure 68 – Transformer Diagnostic Test Chart
(Adapted from IEEE 62-1995™ [20]).

Appendix: Hydroplant Risk Assessment – Transformer Condition Assessment

1. General

Power transformers are key components in the power train at hydroelectric powerplants and are appropriate for analysis under a risk assessment program. Transformer failure can have a significant economic impact due to long lead times in procurement, manufacturing, and installation, in addition to high equipment cost. According to the Electric Power Research Institute (EPRI): "Extending the useful life of power transformers is the single most important strategy for increasing life of power transmission and distribution infrastructures starting with generator step-up transformers (GSU) at the powerplant itself" (EPRI Report No. 1001938).

Determining the existing condition of power transformers is an essential step in analyzing the risk of failure. This chapter provides a process for developing at a Transformer Condition Index. This condition index may be used as an input to the risk-and-economic analysis computer model where it adjusts transformer life expectancy curves. The output of the economic analysis is a set of alternative scenarios, including costs and benefits, intended for management decisions on replacement or rehabilitation.

2. Scope/Application

The transformer condition assessment methodology outlined in this chapter applies to oil-filled power transformers (> 500 kilovoltampere [kVA]) currently in operation.

This guide is not intended to define transformer maintenance practices or describe transformer condition assessment inspections, tests, or measurements in detail. Utility maintenance policies and procedures must be consulted for such information.

3. Condition Indicators and Transformer Condition Index

This guide describes four Condition Indicators generally regarded by hydro powerplant engineers as a sound basis for assessing transformer condition.

♦ Insulating oil analysis (dissolved gas analysis [DGA] and furan)

♦ Power factor and excitation current tests

♦ Operation and maintenance (O&M) history

♦ Age

These indicators are based on "Tier 1" inspections, tests, and measurements conducted by utility staff or contractors over the course of time. The indicators are expressed in numerical terms and are used to arrive at an overall Transformer Condition Index.

The guide also describes a "toolbox" of Tier 2 inspections, tests, and measurements that may be applied, depending on the specific problem being addressed. Results of Tier 2 may modify the score of the Transformer Condition Index.

After review by a transformer expert, the Transformer Condition Index is suitable for use as an input to the risk and economic analysis model.

4. Inspections, Testing, and Measurements

The hierarchy of inspections, tests, and measurements is illustrated in figure A-1, Transformer Condition Assessment Methodology. Table A-14, Transformer Condition Assessment Summary (later in this appendix), summarizes these activities.

Inspections, tests, and measurements ("tests") performed to determine transformer condition are divided into two tiers (or levels). Tier 1 tests are those that are routinely accomplished as part of normal O&M or are readily discernible by examination of existing data. Results of Tier 1 tests are quantified below as Condition Indicator Scores that are weighted and summed to arrive at a Transformer Condition Index. Tier 1 tests may indicate abnormal conditions that can be resolved with standard corrective maintenance solutions. Tier 1 test results may also indicate the need for additional investigation, categorized as Tier 2 tests.

Tier 2 tests are considered nonroutine. Tier 2 test results may affect the Transformer Condition Index established using Tier 1 tests but may confirm also or disprove the need for more extensive maintenance, rehabilitation, or transformer replacement.

Inspection, testing, and measurement methods are specified in technical references specific to the electric utility.

This guide assumes that Tier 1 and Tier 2 inspections, tests, and measurements are conducted and analyzed by staff suitably trained and experienced in transformer diagnostics. More basic tests may be performed by qualified staff who are competent in these routine procedures. More complex inspections and measurements may require a transformer diagnostics "expert."

This guide also assumes that inspections, tests, and measurements are conducted on a frequency that provides accurate and current

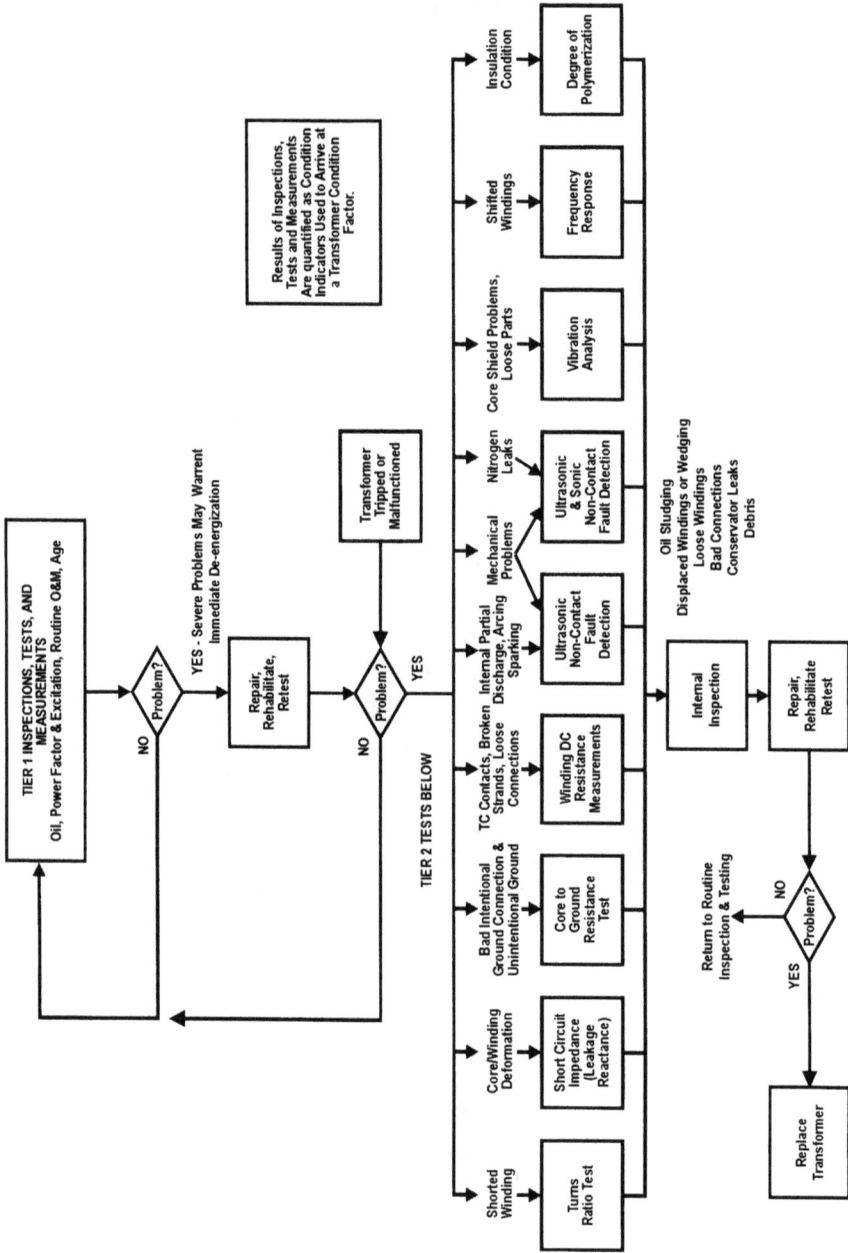

Figure A-1 – Transformer Condition Assessment Methodology.

information needed by the assessment. In some cases, it may be necessary to conduct tests prior to this assessment to acquire current data.

Transformer condition assessment may cause concern that justifies more frequent monitoring. Utilities should consider the possibility of taking more frequent measurements (e.g., oil samples) or installing online monitoring systems (e.g., gas-in-oil) that will continuously track critical quantities. This will provide additional data for condition assessment and establish a certain amount of reassurance as transformer alternatives are being explored.

NOTE:

A severely negative result of ANY inspection, test, or measurement may be adequate, in itself, to require immediate de-energization or prevent re-energization of the transformer, regardless of the Transformer Condition Index score.

---

5. Scoring

Transformer Condition Indicator scoring is somewhat subjective, relying on transformer condition experts. Relative terms such as "results normal" and "degradation" refer to results that are compared to:

♦ Industry accepted levels

♦ Baseline or previous (acceptable) levels on this equipment

♦ Equipment of similar design, construction, or age that operates in a similar environment

6. Weighting Factors

Weighting factors used in the condition assessment methodology recognize that some Condition Indicators affect the Transformer Condition Index to a greater or lesser degree than other indicators.

These weighting factors were arrived at by consensus among transformer design and maintenance personnel with extensive experience.

7. Mitigating Factors

Every transformer is unique; therefore, the methodology described in this chapter cannot quantify all factors that affect individual transformer condition. It is important that the Transformer Condition Index be scrutinized by engineering experts. Mitigating factors specific to the utility may determine the final Transformer Condition Index and the final decision on transformer replacement or rehabilitation.

8. Documentation

Substantiating documentation is essential to support findings of the assessment, particularly where a Tier 1 Condition Indicator score is less than 3 or a Tier 2 test results in subtractions from the Transformer Condition Index. Test results and reports, photographs, O&M records, or other documentation should accompany the Transformer Condition Assessment Summary Form.

9. Condition Assessment Methodology

The condition assessment methodology consists of analyzing each Condition Indicator individually to arrive at a Condition Indicator Score; then the score is weighted and summed with scores from other condition indicators. The sum is the Transformer Condition Index. Apply the Condition Index to the Alternatives table to determine the recommended course of action.

Reasonable efforts should be made to perform Tier 1 inspections, tests, and measurements. However, when data is missing, to properly score the Condition Indicator, it may be assumed that the score is "Good" or a mid-range number such as 2.

CAUTION:

This strategy should be used judiciously to prevent deceptive results.

---

10. Tier 1 – Inspections, Tests, Measurements

Tier 1 inspections, tests, and measurements routinely accomplished as part of normal O&M are readily discernible by examination of existing data. Tier 1 test results are quantified below as Condition Indicators that are weighted and summed to arrive at a Transformer Condition Index. Tier 1 inspections, tests, and measurements may indicate abnormal conditions that can be resolved with standard corrective maintenance solutions. Tier 1 test results may also indicate the need for additional investigation, categorized as Tier 2 tests.

10.1 Condition Indicator 1 – Insulating Oil Analysis

Dissolved gas analysis is the most important factor in determining the condition of a transformer because, being performed more frequently than other tests, it may be the first indication of a problem. Insulating oil analysis can identify internal arcing, bad electrical contacts, hot spots, partial discharge, or overheating of conductors, oil, tank, or cellulose. The "health" of the oil is reflective of the health of the transformer itself. DGA consists of sending transformer insulating oil samples to a commercial laboratory for analysis. The most important indicator is the individual and total dissolved combustible gas (TDCG) generation rates, based on International Electrotechnical Commission (IEC) and Institute of Electrical and Electronic Engineers (IEEE®) standards. Although gas generation rates are not the only indicator, they are reasonable for use in determining the Condition Indicator Score.

Furanic analysis may indicate a problem with the paper insulation, which could affect transformer longevity. A baseline furanic analysis should be made initially and repeated if the transformer is overheated, overloaded, aged, or after changing or processing the oil.

Physical tests such as interfacial tension (IFT), acidity, moisture content, and dielectric strength usually indicate oil conditions that can be remedied through various reclamation processes. Therefore, they are not indicative of overall transformer condition that would lead to replacement. Such tests do not affect the Insulating Oil Condition Indicator score.

Results are analyzed and applied to table A-1 to arrive at a Condition Indicator Score.

NOTE:

Choose one furan number from table A-1, based on a transformer rated 65 °C or 55 °C rise above ambient temperature.

The below DGA numbers are based on dissolved gas in oil generation rates and come from a combination of IEEE C57-104™, IEC 60599, and Delta X Research's Transformer Oil Analysis (TOA) software.

Table A-1 – Insulating Oil Analysis Scoring

Results	Condition Indicator Score
TDCG generation rate less than 30 parts per million per month (ppm/month) and: * All individual combustible gas generation rates less than 10 ppm/month * Exceptions: carbon monoxide (CO) generation rate less than 70 ppm/month and acetylene (C_2H_2) generation rate 0 ppm. (See Note below.) *AND* 2FAL furans 150 parts per billion (ppb) or less for 65 °C rise. 2FAL furans 450 ppb or less for 55 °C rise	3
TDCG generation rate between 30 and 60 ppm/month and: * All individual combustible gas generation rates less than 15 ppm/month. * Exceptions: CO generation rate less than 150 ppm/month and C_2H_2 generation rate 0 ppm. (See Note below.) *OR* 2FAL furans between 150 ppb and 200 ppb for 65 °C rise. 2FAL furans between 450 ppb and 600 ppb for 55 °C rise.	2
TDCG generation rate between 50 and 80 ppm/month and: * All individual combustible gas generation rates less than 25 ppm/month. * Exceptions: CO generation rate less than 350 ppm/month and C_2H_2 generation rate less than 5 ppm/month. (See Note below.) *OR* 2FAL furans between 200 ppb and 250 ppb for 65 °C rise. 2FAL furans between 600 ppb and 750 ppb for 55 °C rise.	1
TDCG generation rate greater than 80 ppm/month and: * Any individual combustible gas generation rate more than 50 ppm/month. * Exceptions: CO generation rate more than 350 ppm/month and C_2H_2 generation rate no more than 10 ppm/per month. (See Note below.) *OR* 2FAL furans above 250 ppb for 65 °C rise. 2FAL furans above 750 ppb for 55 °C rise.	0

213

NOTE:

Any ongoing acetylene (C_2H_2) generation indicates an active arcing fault, and the transformer may have to be removed from service to avoid possible catastrophic failure. A transformer may be safely operated with some C_2H_2 showing in the DGA. C_2H_2 sometimes comes from a one-time event such as a close-in lightning strike or through fault. However, if C_2H_2 is increasing more than 10 ppm per month, the transformer should be removed from service. Because acetylene generation is a critical indicator of transformer internal condition, each utility should establish practices in accordance with published standards and transformer experts to monitor any increases in gas generation and take corrective action. Increasing the frequency of DGA and de-gasifying the transformer oil are potential alternatives to consider.

10.2 Condition Indicator 2 – Power Factor and Excitation Current Tests

Power factor insulation testing is important to determining the condition of the transformer because it can detect winding and bushing insulation integrity. Power factor and excitation current tests are conducted in the field on de-energized, isolated, and properly grounded transformers. Excitation current tests measure the single-phase voltage, current, and phase angle between them, typically on the high-voltage side with the terminals of the other winding left floating (with the exception of a grounded neutral). The measurements are performed at rated frequency and usually at test voltages up to 10 kilovolts (kV). The test detects shorted turns, poor tap changer contacts, and core problems.

Results are analyzed and applied to table A-2 to arrive at a Condition Indicator Score.

Table A-2.—Power Factor and Excitation Current Test Scoring

Test Results[1]	Condition Indicator Score
Power factor results normal (Good - G). *AND* Normal excitation current values and patterns, compared to other phases and prior tests.	*3*
Power factor results show minor degradation. (Deteriorated - D). *OR* Minor deviation[2] in excitation current values and patterns, compared to other phases and prior tests.	*2*
Power factor results show significant deterioration. (Investigate - I) *OR* Significant deviation[2] in current values and patterns, compared to other phases and prior tests.	*1*
Power factor results show severe degradation. (Bad – B) *OR* Severe deviation[2] in current values and patterns, compared to other phases and prior tests.	May indicate serious problem requiring immediate evaluation, additional testing, consultation with experts, and remediation prior to re-energization.

[1] Doble insulation rating in parentheses.
[2] Be sure to account for residual magnetism and load tap changer (LTC).

10.3 Condition Indicator 3 – Operation and Maintenance History

O&M history may indicate overall transformer condition. O&M history factors that may apply are:

♦ Sustained overloading

♦ Unusual operating temperatures indicated by gauges and continuous monitoring

♦ Abnormal temperatures indicated by infrared scanning

♦ Nearby lightning strikes or through faults

♦ Abnormally high corona detected.

♦ Abnormally high external temperatures detected.

♦ Problems with auxiliary systems (fans, radiators, cooling water piping, pumps, motors, controls, nitrogen replenishment system, and indicating and protection devices).

♦ Deteriorated control and protection wiring and devices.

♦ Increase in corrective maintenance or difficulty in acquiring spare parts.

♦ Anomalies determined by physical inspection (external inspection or internal inspection not requiring untanking) (e.g., incorrectly positioned valves, plugged radiators, stuck temperature indicators and level gauges, noisy oil pumps or fans, oil leaks, connections to bushings).

♦ Previous failures on this equipment.

♦ Failures or problems on equipment of similar design, construction, or age that operate in a similar environment.

Qualified personnel should make a subjective determination of scoring that encompasses as many O&M factors as possible under this indicator.

Results are analyzed and applied to table A-3 to arrive at a Condition
Indicator Score.

Table A-3 – Operation and Maintenance History Scoring

History Results	Condition Indicator Score
Operation and maintenance are normal.	3
Some abnormal operating conditions experienced and/or additional maintenance above normal are occurring.	2
Significant operation outside normal and/or significant additional maintenance is required; or forced outage occurs; or outages are regularly extended due to maintenance problems; or similar units are problematic.	1
Repeated forced outages; maintenance not cost effective; or major oil leaks and/or severe mechanical problems; or similar units have failed.	0

10.4 Condition Indicator 4 – Age

Transformer age is an important factor to consider when identifying
candidates for transformer replacement. Age is one indicator of
remaining life and upgrade potential to current state-of-the-art
materials. During the life of the transformer, the structural and
insulating properties of materials used for structural support and
electrical insulation, especially wood and paper, deteriorate. Although
actual service life varies widely depending on the manufacturer's
design, quality of assembly, materials used, operating history, current
operating conditions, and maintenance history, the average expected
life for an individual transformer in a large population of transformer
is statistically about 40 years.

Apply the transformer age to table A-4 to arrive at the Condition
Indicator Score.

Table A-4 – Age Scoring

Age	Condition Indicator Score
Under 30 years	3
30 – 45 years	2
Over 45 years	1

11. Tier 1 – Transformer Condition Index Calculations

Enter the Condition Indicator Scores from the tables above into the Transformer Condition Assessment Summary Form at the end of this appendix. Multiply each Condition Indicator Score by the Weighting Factor, then sum the Total Scores to arrive at the Tier 1 Transformer Condition Index. This index may be adjusted by the Tier 2 inspections, tests, and measurements described below. Suggested alternatives for follow-up action, based on the Transformer Condition Index, are described in table A-14 at the end of this appendix.

12. Tier 2 – Inspections, Tests, Measurements

Tier 2 inspections, tests, and measurements generally require specialized equipment or training, may be intrusive, or may require an extended outage to perform. Tier 2 assessment is considered nonroutine. Tier 2 inspections may affect the Transformer Condition Index number established using Tier 1 but also may confirm or refute the need for more extensive maintenance, rehabilitation, or transformer replacement.

For each Tier 2 inspection, test, or measurement performed, subtract the appropriate amount from the appropriate Tier 1 Condition ndicator and recalculate the Transformer Condition Index using the Transformer Condition Assessment Summary Form at the end of this document.

12.1 Test T2.1: Turns Ratio Test

The transformer turns ratio (TTR) test detects shorts between turns of he same coil, which indicates insulation failure between the turns. These tests are performed with the transformer de-energized and may show the necessity for an internal inspection or removal from service.

Results are analyzed and applied to table A-5 to arrive at a Transformer Condition Index Score adjustment.

Table A-5 – Turns Ratio Test Scoring

Test Results	Adjustment to Transformer Condition Index
Less than 0.20% difference from nameplate V_1/V_2 and compared to previous readings	No change
0.20% to 0.50% difference compared to nameplate V_1/V_2	Subtract 1.0
Greater than 0.5% difference compared to nameplate V_1/V_2	May indicate serious problem requiring immediate evaluation, additional testing, consultation with experts, and remediation prior to re-energization.

12.2 Test T2.2: Short Circuit Impedance Tests

Sometimes called percent impedance or leakage reactance test, these tests are conducted in the field and compared to nameplate information, previous tests, and similar units to detect deformation of the core or windings caused by shipping damage, through faults, or ground faults. Some difference may be expected between nameplate and field test results because factory tests are conducted at full load current, which is normally not possible in the field. Field connections and test leads and jumpers also play a significant role in test results, and it is impossible to exactly duplicate the factory test setup.

Therefore, the I^2R losses may be different and cause different test results. By comparing percent-reactance to nameplate impedance, the differences caused by leads and connections can be eliminated. Because reactance is only the inductive component of the impedance, I^2R losses are omitted in the test results.

Results are analyzed and applied to table A-6 to arrive at a Transformer Condition Index Score adjustment.

Table A-6 – Short Circuit Test Scoring

Test Results	Condition Indicator Score
Less than 1% difference from nameplate impedance	No change
1% to 3% difference from nameplate impedance (minor degradation)	Subtract 0.5
3% to 5% difference from nameplate impedance (significant degradation)	Subtract 1.0
Greater than 5% difference from nameplate impedance (severe degradation)	May indicate serious problem requiring immediate evaluation, additional testing, consultation with experts, and remediation prior to re-energization.

12.3 Test T2.3: Core-to-Ground Resistance Megger® Tests

The transformer core is intentionally grounded through one connection. The core-to-ground resistance test can detect if this connection is loose. It can also detect whether there are other undesired and inadvertent, grounds. If the intentional core ground is intact, the resultant resistance should be very low. To check for unintentional core grounds, remove the intentional ground and Megger® between the core and the grounded transformer tank. This test should produce very high resistance, indicating that an unintentional ground is not present. This test is to supplement DGA that shows generation of hot metal gases (methane, ethane, and ethylene) and to indicate if a spurious, unintentional core ground is the problem. Experience can help locate the source of the problem.

Results are analyzed and applied to table A-7 to arrive at a Transformer Condition Index score adjustment.

Table A-7 – Core-to-Ground Resistance Test Scoring

Test Results[1]	Adjustment to Transformer Condition Index
Greater than 1,000 megohms[1] (results normal)	3
600 to 900 megohms	2
200 to 500 megohms	1
Less than 200 megohms	May indicate serious problem requiring immediate evaluation, additional testing, consultation with experts, and remediation prior to re-energization.

[1] With intentional ground disconnected.

12.4 Test T2.4: Winding Direct-Current Resistance Measurements

Careful measurement of winding resistance can detect broken conductor strands, loose connections, and bad contacts in the tap changer (de-energized tap changer [DETC] or LTC). Results from these measurements may indicate the need for an internal inspection. This information supplements DGA and is useful when DGA shows generation of heat gases (ethane, ethylene, and methane). These tests are typically performed with a micro-ohmmeter and/or a Wheatstone bridge. Test results are compared between phases or with factory tests. When comparing with factory tests, a temperature correction must be employed (IEEE P62™). This test should be performed only after the rest of the routine electrical tests because it may magnetize the core, affecting results of the other tests.

Results are analyzed and applied to table A-8 to arrive at a Transformer Condition Index score adjustment.

12.5 Test T2.5: Ultrasonic and Sonic Fault Detection Measurements

These assessment tests (sometimes called acoustic testing tests) are helpful in locating internal faults. Partial discharges (corona) and low-energy arcing/sparking emit energy in the range of 50 megahertz (ultrasonic), well above audible sound. To make these measurements, sensors are placed on the outside of a transformer tank to detect these ultrasonic emissions, which are then converted electronically to oscilloscope traces or audible frequencies and recorded. By triangulation, a general location of a fault (corona or arcing/sparking) may be determined so that an internal inspection can be focused in that location. These devices also can detect loose shields that build up static and discharge it to the grounded tank, poor connections on bushings, bad contacts on a tap changer that are arcing/sparking, core ground problems that cause sparking/arcing, and areas of weak insulation that generate corona. Sonic testing can detect increased core and coil noise (looseness) and vibration, failing bearings in oil pumps and fans, and nitrogen leaks in nitrogen blanketed transformers.

Table A-8 – Winding Direct-Current Resistance Measurement Scoring

Measurement Results	Adjustment to Transformer Condition Index
No more than 5% difference between phases or from factory tests	No change
5% to 7% difference between phases or from factory tests	Subtract 0.5
7% to 10% difference between phases or from factory tests	Subtract 1.0
More than 10% difference between phases or from factory tests	May indicate serious problem requiring immediate evaluation, additional testing, consultation with experts, and remediation prior to re-energization.

Information gained from these measurements supplements DGA and provides additional support information for de-energized tests, such as core ground and winding resistance tests. In addition, these tests help pinpoint areas to look for problems during internal inspections.

Performing baseline tests may provide comparisons for later tests. Experience can help locate the source of the problem.

Results are analyzed and applied to table A-9 to arrive at a Transformer Condition Index score adjustment.

Table A-9 – Ultrasonic and Sonic Measurement Scoring

Measurement Results	Adjustment to Transformer Condition Index
Results normal	No change
Low-level fault indication	Subtract 0.5
Moderate level fault indication	Subtract 1.0
Severe fault level indication	Subtract 2.0

12.6 Test T2.6: Vibration Analysis

Vibration can result from loose transformer core and coil segments, shield problems, loose parts, or bad bearings on oil cooling pumps or fans. Vibration analyzers are used to detect and measure the vibration. Information gained from these tests supplements ultrasonic and sonic (acoustic) fault detection tests and DGA. Information from these tests may indicate maintenance is needed on pumps/fans mounted external to the tank. It may also show when an internal transformer inspection is necessary. If wedging has been displaced due to paper deterioration or through faults, vibration will increase markedly. This will also show if core and coil vibration has increased compared to baseline information. Experience can help locate the source of the problem.

Results are analyzed and applied to table A-10 to arrive at a
Transformer Condition Index score adjustment.

Table A-10 – Vibration Analysis Scoring

Analysis Results	Adjustment to Transformer Condition Index
Results normal	No change
Low level fault indication	Subtract 0.5
Moderate level fault indication	Subtract 1.0
Severe fault level indication	Subtract 2.0

12.7 Test T2.7: Frequency Response Analysis (FRA)

Frequency Response Analysis (or Sweep Frequency Response
Analysis) can determine if windings of a transformer have moved or
shifted. It can be completed as a factory test prior to shipment and
repeated after the transformer is received onsite to determine if
windings have been damaged or shifted during shipping. This test is
also helpful if a protective relay has tripped or a through fault, short
circuit, or ground fault has occurred

A sweep frequency test signal is generally placed on each of the high-
voltage windings, and the signal is detected on the low-voltage
windings. This provides a picture of the frequency transfer function of
the windings. If the windings have been displaced or shifted, test
results will differ markedly from prior tests. Test results are kept in
transformer history files so they can be compared to later tests.
Results are determined by comparison to baseline or previous
measurements or by comparison to units of similar design and
construction.

Results are analyzed and applied to table A-11 to arrive at a
Transformer Condition Index Score adjustment.

Table A-11 – Frequency Response Analysis Scoring

Test Results	Adjustment to Transformer Condition Index
No deviation compared to prior tests	No change
Minor deviation compared to prior tests	Subtract 2.0
Moderate deviation compared to prior tests	Subtract 3.0
Significant deviation compared to prior tests	Subtract 4.0
Severe deviation compared to prior tests	May indicate serious problem requiring immediate evaluation, additional testing, consultation with experts, and remediation prior to re-energization.

12.8 Test T2.8: Internal Inspection

In some cases, it is necessary to open, partially or fully drain the oil, and perform an internal inspection to determine the transformer's condition. These inspections must be performed by experienced staff with proper training. Sludging, loose wedges, loose coils, poor electrical connections on bushing bottoms, burned contacts on tap changers, localized overheating signified by carbon buildup, displaced wedging or insulation, and debris and other foreign material are general matters of concern. Photographs and mapping problem locations are good means of documenting findings. Before entering the transformer, and while inside it, the Occupational Safety and Health Administration, State, local, and utility safety practices must be followed (e.g., "permitted confined space" entry practices).

Results are analyzed and applied to table A-12 to arrive at a Transformer Condition Index Score adjustment.

12.9 Test T2.9: Degree of Polymerization

Winding insulation (cellulose) deterioration can be quantified by analysis of the degree of polymerization (DP) of the insulating material. This test gives an indication of the remaining structural

Table A-12 – Internal Inspection Scoring

Inspection Results	Adjustment to Transformer Condition Index
Conditions normal	No change
Minimal degradation	Subtract 0.5
Moderate degradation	Subtract 1.5
Severe degradation	May indicate serious problem requiring immediate evaluation, additional testing, consultation with experts, and remediation prior to re-energization.

strength of the paper insulation and is an excellent indication of the remaining life of the paper and the transformer itself. This requires analyzing a sample of the paper insulation in a laboratory to determine the deterioration of the molecular bonds of the paper.

Results are analyzed and applied to table A-13 to arrive at a Transformer Condition Index score adjustment.

Table A-13 – Degree of Polymerization Scoring

Test Results	Adjustment to Transformer Condition Index
900 or higher (no polymerization decrease; results normal)	No change
800-700 (minimal polymerization decrease)	Subtract 0.5
600-300 (moderate polymerization decrease)	Subtract 1.5
<200 (severe polymerization decrease— insulation has no mechanical strength; end of life)	May indicate serious problem requiring immediate evaluation, additional testing, consultation with experts, and remediation prior to re-energization.

13. Tier 2 – Transformer Condition Index Calculations

Enter the Tier 2 adjustments from the tables above into the Transformer Condition Assessment Summary Form at the end of this chapter. Subtract the sum of these adjustments from the Tier 1 Transformer Condition Index to arrive at the total Transformer Condition Index. Suggested alternatives for followup action, based on the Transformer Condition Index, are described in table A-14.

Table A-14 – Transformer Co-ndition Assessment Summary

Test	Detects	Tool
	Online Tests	
Dissolved Gas Analysis	Internal arcing; bad electrical contacts; hot spots; partial discharge; and overheating of conductors, oil, tank, and cellulose insulation.	**Requires laboratory** analysis
Oil Physical and Chemical Tests	Moisture, degraded IFT, acidity, furans, dielectric strength, and power factor.	**Requires laboratory** analysis
External Physical Inspection	Oil leaks, broken parts, worn paint, defective support structure, stuck indicators, noisy operation, loose connections, cooling problems with fans, pumps, etc.	**Experienced staff** and binoculars
External Temperatures (Main tank and load tap changer)	Temperature monitoring with changes in load and ambient temperature.	**Portable temperature** data loggers and software
Infrared Scan	Hot spots indicating localized heating, circulating currents, blocked cooling, tap changer problems, and loose connections.	**Thermographic** camera and analysis software
Ultrasonic (Acoustic) Contact Fault Detection	Internal partial discharge, arcing, sparking, loose shields, poor bushing connections, bad tap changer contacts, core ground problems, and weak insulation that is causing corona.	**Ultrasonic detectors** and analysis software
Sonic Fault Detection	Nitrogen leaks, vacuum leaks, core and coil vibration, corona at bushings, and mechanical and bearing problems in pumps and cooling fans.	**Ultrasonic probe** and meter
Vibration Analysis	Internal core, coil, and shield problems; loose parts and bad bearings.	Vibration data logger

227

Table A-14 – Transformer Condition Assessment Summary (continued)

Test	Detects	Tool
	Offline Tests	
Doble Tests (bushing capacitance, insulation power factor, tip up, excitation current)	Loss of winding insulation integrity, loss of bushing insulation integrity, and winding moisture.	Doble test equipment
Turns Ratio	Shorted windings.	Doble test equipment or turns ratio tester
Short Circuit Impedance	Deformation of the core or winding.	Doble or equivalent test equipment
Core to Ground Resistance (External test may be possible depending on transformer construction)	Bad connection on intentional core ground and existence of unintentional grounds.	Megger®
Winding Direct Current Resistance Measurements	Broken strands, loose connections, bad tap changer contacts.	Wheatstone Bridge (1 ohm and greater) Kelvin Bridge micro-ohmmeter (less than 1 ohm)
Frequency Response Analysis	Shifted windings.	Doble or equivalent sweep frequency analyzer
Internal Inspection	Oil sludging, displaced winding or wedging, loose windings, bad connections, localized heating, and debris and foreign objects.	Experienced staff, micro-ohmmeter
Degree of Polymerization	Insulation condition (life expectancy).	Laboratory analysis of paper sample

TRANSFORMER CONDITION ASSESSMENT
SUMMARY FORM

Date: _____   Transformer Identifier: _____
Location: _____   Manufacturer: _____ Yr. Mfd: _____
No. of Phases: _____   MVA: _____   Voltage: _____

Tier 1 Transformer Condition Summary			
Indicator No.	**Indicator**	**Score X Weighting Factor = Total Score**	
1	Oil Analysis	1.143	
2	Power Factor and Excitation Current Tests	0.952	
3	Operation and Maintenance History	0.762	
4	Age	0.476	
	Tier 1 Transformer Condition Index (sum of individual Total Scores)		(Between 0 and 10)

Tier 2 Transformer Condition Summary	
Tier 2 Test	**Adjustment to Tier 1 Condition Index**
T2.1 Turns Ratio Test	
T2.2 Short Circuit Impedance Test	
T2.3 Core-to-Ground Resistance Megger® Test	
T2.4 Winding Direct Current Resistance Measurement	
T2.5 Ultrasonic and Sonic Fault Detection Measurement	
T2.6 Vibration Analysis	
T2.7 Frequency Response Analysis	
T2.8 Internal Inspection	
T2.9 Degree of Polymerization	
Tier 2 Adjustments to Transformer Condition Index (Sum of individual adjustments)	

To calculate the net Transformer Condition Index, subtract the Tier 2 Adjustments from the Tier 1 Condition Index:

Tier 1 Transformer Condition Index _____

minus   **Tier 2 Adjustments** _____ = _____

                                           Net Transformer Condition Index

Evaluator: _____

Technical Review: _____   Management Review: _____

Copies to: _____

Forward completed form, including supporting documentation, to agency transformer expert.

14. Transformer Alternatives

After review by a transformer expert, the Transformer Condition Index—either modified by Tier 2 tests or not—may be sufficient for decision-making regarding transformer alternatives. The index is also suitable for use in the risk-and-economic analysis model described elsewhere. Where it is desired to consider alternatives based solely on transformer condition, the Transformer Condition Index may be directly applied to table A-15.

Table A-15 – Transformer Condition-Based Alternatives

Transformer Condition Index Range	Suggested Course of Action
8-10 (Good)	Continue O&M without restriction. Repeat this condition assessment process as needed.
4-7 (Fair)	Continue operation but re-evaluate O&M practices. Consider using appropriate Tier 2 tests. Conduct full Life Extension risk-economic assessment. Repeat this condition assessment process as needed.
0-3 (Poor)	Immediate evaluation including additional Tier 2 testing. Consultation with experts. Adjust O&M as prudent. Begin replacement/rehabilitation process.

References

[1] *Standard General Requirements for Dry-Type Distribution, Power, and Regulating Transformers,* American National Standards Institute/Institute of Electrical and Electronic Engineers (ANSI/IEEE) C57.12.01, 1989.

[2] Standard General Requirements for Liquid-Immersed Distribution, Power and Regulation Transformers, Institute of Electrical and Electronic Engineers (IEEE) C57.12.00, 1993.

[3] Power Transformer Maintenance and Testing, General Physics Corporation, 1990.

[4] *Guidelines for the Life Extension of Substations,* Electric Power Research Institute (EPRI) TR-105070, April 1995.

[5] *Guidelines for the Life Extension of Substations,* Electric Power Research Institute (EPRI), 2002 update.

[6] *Transformer Maintenance Guide,* J.J. Kelly, S.D. Myers, R.H. Parrish, S.D. Myers Co., 1981.

[7] Transformer General Gasketing Procedures, Alan Cote, S.D. Myers Co., 1987.

[8] Recommended Practice for Electrical Equipment Maintenance, National Fire Protection Association (NFPA) 70-B-2002.

[9] *Bushing Field Test Guide,* Doble Engineering Company, 1966.

[10] *Testing and Maintenance of High-Voltage Bushings, FIST 3-2,* Bureau of Reclamation, 1991.

[11] General Requirements and Test Procedure for Outdoor Power Apparatus Bushings, Institute of Electrical and Electronic Engineers (IEEE) C57.10.00, 1991.

[12] *Guide for the Interpretation of Gases Generated in Oil-Immersed Transformers*, Institute of Electrical and Electronic Engineers (IEEE) C57.104, 1991.

[13] "Mineral Oil-Impregnated Electrical Equipment" in *Service-Interpretation of Dissolved and Free Gas Analysis,* International Electrotechnical Commission (IEC) 60599, 1997.

[14] *Dissolved Gas Analysis of Transformer Oil,* John C. Drotos, John W. Porter, Randy Stebbins, S.D. Myers Co., 1996.

[15] Recommended Practice for Installation, Application, Operation and Maintenance of Dry-Type General Purpose Distribution and Power Transformers, Institute of Electrical and Electronic Engineers (IEEE) C57.94, 1982.

[16] Criteria for the Interpretation of Data for Disssolved Gases in Oil from Transformers (A Review), by Paul Griffin, Doble Engineering Co., 1996.

[17] *Maintenance of High Voltage Transformers,* Martin Heath Cote Associates, London, England, 1989.

[18] *Thermal Monitors and Loading,* Harold Moore, Transformer Performance Montitoring and Diagnostics EPRI, September 1997.

[19] IEEE and IEC Codes to Interpret Incipient Faults in Transformers, Using Gas in Oil Analysis, R.R. Rogers C.E.G.B, Transmission Division, Guilford, England, circa 1995.

[20] IEEE Guide for Diagnostic Field Testing of Electrical Power Apparatus, Part 1: Oil Filled Power Transformers, Regulators, and Reactors, Institute of Electrical and Electronic Engineers (IEEE) 62, 1995.

[21] Maintenance of Liquid Insulation: Mineral Oils and Askarels, FIST 3-5, Bureau of Reclamation, 1992.

[22] *Standard Test Method for Interfacial Tension of Oil Against Water by the Ring Method,* American National Standards Institute/American Society for Testing and Materials (ANSI/ASTM) D 971-91.

[23] *Experience with In-Field Water Contamination of Large Power Transformers,* Victor V. Sokolov and Boris V. Vanin, Scientific and Engineering Center "ZTZ Service Co.," EPRI Substation Equipment Diagnostics Conference VII, Ukraine, 1999.

[24] Proceedings: Substation Equipment Diagnostics, Conference IX, Electric Power Research Institute (EPRI), 2000.

[25] *Reference Book for Insulating Liquids and Gases,* Doble Engineering Company RBILG-391, 1993

[26] *Guide for Loading Mineral Oil Immersed Transformers,* American National Standards Institute/Institute of Electrical and Electronic Engineers (ANSI/IEEE) C57.92, 1981.

[27] *Insulating Fluids,* Doble Engineering Company Client Conference Minutes No. 65PAIC98, 1998.

[28] Trial-Use Guide for the Interpretation of Gases Generated in Silicone-Immersed Transformers, Institute of Electrical and Electronic Engineers (IEEE) P1258, 1999.

[29] Standard Specification of Nitrogen Gas as an Electrical Insulating Material, American Society for Testing and Materials (ASTM) D-1933-97.

[30] Standard Specification for Mineral Insulating Oil Used in Electrical Apparatus, American Society for Testing and Materials (ASTM) D-3487, 1988.

[31] Standard Test Method for Furanic Compounds in Electrical Insulating Liquids by High Performance Liquid Chromatography, American Society for Testing and Materials (ASTM) D-5837, 1996.

[32] Standard Test Method for Compressibility and Recovery of Gasket Materials, American Society of Testing and Materials (ASTM) F-36, 1999.

[33] *Standard Test Method for Rubber Property – Durometer Hardness,* American Society of Testing and Materials (ASTM) D-2240, 1997.

[34] "Transformer Condition Assessment," in *Hydro Powerplant Risk Assessment Guide,* a joint effort of Bureau of Reclamation, U.S. Army Corps of Engineers, Hydro Quebec, and Bonneville Power Administration (see appendix).

[35] *Replacements,* Bureau of Reclamation and Western Area Power Administration, July 1995.

[36] M5100 Sweep Frequency Response Analysis (SFRA) Instrument Users Guide, Doble Engineering Company, 2001.

[37] Maintenance Scheduling for Electrical Equipment, FIST 4-1B, Bureau of Reclamation, 2001.

[38] Thermographic Maintenance Program, FIST 4-13, Bureau of Reclamation.

[39] *An Introduction to the Half-Century Transformer,* Transformer Maintenance Institute, S.D. Myers, Co., 2002.

[40] Transformer Nitrogen Advisory, PEB 5, October 2000.

Acronyms and Abbreviations

AC	alternating current
amp	ampere
ANSI	American National Standards Institute
ASTM	American Society for Testing and Materials
back EMF	back electro-motive force
cfm	cubic feet per minute
CH$_4$	methane
C$_2$H$_2$	acetylene
C$_2$H$_4$	ethylene
C$_2$H$_6$	ethane
CO	carbon monoxide
CO$_2$	carbon dioxide
CT	current transformer
dB	decibel
DBPC	ditertiary butyl paracresol (oxygen inhibitor for transformer oil)
dc	direct current
DETC	de-energized tap changer
DGA	dissolved gas analysis
DP	degree of polymerization (measure of mechanical strength and remaining life of paper insulation).
EHV	extra high voltage (higher than 240,000 volts)
EPRI	Electric Power Research Institute
FIST	Facilities Instructions, Standards, and Techniques
FRA	Frequency Response Analyses

gm	gram
GSU	generator step-up
H2	hydrogen
Hz	hertz (cycles per second)
IEC	International Electrotechnical Commission
IEEE	Institute of Electrical and Electronic Engineers
IFT	Interfacial Tension (between transformer oil and water; this is a measure of the amount of particles and pollution in transformer oil)
ips	inches per second
IR	infrared
I²R	watts
JHA	job hazard analysis
kHz	kilohertz (kilocycles per second)
KOH	potassium hydroxide (used to determine acid number of transformer oil)
kV	kilovolts
kW	kilowatt
LTC	load tap changer
kVA	kilovoltampere
mA	milliampere
M/DW	moisture by dry weight (percent of weight of water in paper insulation based on dry weight of paper given on transformer nameplate)
Mg	milligram
MHz	megahertz (megacycles per second)
MVA	megavoltampere
NFPA	National Fire Protection Association

N$_2$	nitrogen
O&M	operation and maintenance
O$_2$	oxygen
ppb	parts per billion
ppm	parts per million
ppm/day	parts per million per day
psi	pounds per square inch
PT	potential transformer
Reclamation	Bureau of Reclamation
SCADA	supervisory control and data acquisition
SFRA	Sweep Frequency Response Analysis
STP	Standard temperature and pressure; 0 °C (32 °F) and 14.7 psi (one atmosphere)
TCG	total combustible gas
TDCG	total dissolved combustible gas
TOA	Transformer Oil Analysis
TSC	Technical Service Center
TTR	transformer turns ratio
UV	ultraviolet
VA	voltampere
°C	degrees Celsius
°F	degrees Fahrenheit
%	percent

www.ingramcontent.com/pod-product-compliance
Lightning Source LLC
Chambersburg PA
CBHW031949180326
41458CB00006B/1666

9 781780 393544